At the **Root** of **Things**

The Subatomic World

At the **Root** of **Things**
The Subatomic World

Palash B. Pal

Saha Institute of Nuclear Physics

Calcutta, India

Translated from Bengali by Sushan Konar

CRC Press

Taylor & Francis Group

Boca Raton London New York

CRC Press is an imprint of the
Taylor & Francis Group, an **informa** business

A CHAPMAN & HALL BOOK

CRC Press
Taylor & Francis Group
6000 Broken Sound Parkway NW, Suite 300
Boca Raton, FL 33487-2742

© 2014 by Taylor & Francis Group, LLC
CRC Press is an imprint of Taylor & Francis Group, an Informa business

No claim to original U.S. Government works

International Standard Book Number-13: 978-1-4665-9129-5 (Paperback)

Library of Congress Cataloging-in-Publication Data

Pal, P. B. (Palash B.), author.
    At the root of things : the subatomic world / Palash Baran Pal.
        pages cm
    Includes bibliographical references and index.
        ISBN 978-1-4665-9129-5 (pbk. : alk. paper) 1. Condensed matter. 2. Quantum theory.
    3. Particles (Nuclear physics) I. Title.

QC173.454.P35 2015
539.7'2--dc23

2014022602

Visit the Taylor & Francis Web site at
http://www.taylorandfrancis.com

and the CRC Press Web site at
http://www.crcpress.com

– Dedication –

To

Syamal K. Das

Contents

Contents

List of Figures

List of Tables

Preface

This book is a translation of my Bengali book *Ki diye somosto-kichu gorha*. This preface, however, is not a translation of the preface of the Bengali book. It has been written specifically for this English edition.

The book is about the structure of matter, that is to say, on the fundamental building blocks that constitute everything in the universe. The story has been divided into five chapters. In Chapter 1, I introduce a summary of classical (or pre-quantum) physics in order to set the stage for later developments in quantum physics, which are essential for the study of elementary particles. I present this summary from a somewhat unusual point of view, viz., keeping a focus on various conservation laws. In Chapter 2, I discuss the emergence of quantum theory from studies on radiation of heat and photoelectric effect, developments that lead to the concept of the duality between waves and particles. Chapter 3 discusses how quantum theory helped us understand the structure of atoms. Chapter 4 starts with the discovery of particles that were not constituents of atoms, for example, the positron and the muon. Then it goes on to reveal the world of dozens and dozens of particles that were discovered experimentally in the middle of the twentieth century, and then describes how the concept of quarks reduced this maze into something simple and tractable, and ends with the quarks and leptons, which are considered the fundamental particles today. Of course all physical phenomena involve interactions between these particles, and in Chapter 5 I introduce the fundamental interactions, describe the basic nature of quantum theories of these interactions, and end with a discussion of how these interactions might be unified.

I believe that no prior knowledge of physics is necessary for understanding the subject matter of this book. A very rudimentary acquaintance with mathematics is necessary for understanding the few mathematical formulas that appear in the book. To be precise, one only needs to know a few pieces of mathematical notation, which have all been summarized on page xxi. However, the main flow of argument can be followed even without paying attention to most of these equations.

The Bengali original of the book was first published in 1987. The present translation mostly follows the second edition, published in 1997. I used the word 'mostly', because a literal translation was not attempted for many reasons. First, the material of the original book, although quite up-to-date at the time it was published, became outdated in places owing to new discoveries that have occurred since then. So I had to have discussion sessions with the translator, to delete a few lines or add a new paragraph here and there to bring the book up to date. Such updating has taken place in only the last few sections of Chapters 4 and 5. Second, the translation gave me a chance to re-evaluate the original, and I was dissatisfied with some lines of the original, sometimes because the translator pointed them out to me and sometimes otherwise. I felt that they should be modified, and so we did just that. Such lines are strewn all over the book, but fortunately there are not many of them.

There is a third reason. A few years ago I was asked to contribute an article to the science magazine *Resonance*. Rather than attempting something completely from scratch, I penned an article based on the fifth chapter of my book *Ki diye...*. In fact, it was little more than a free translation of that chapter. In an accompanying note, I mentioned that I had made conscious deviations only in places where the chapter referred to earlier chapters in the book, and in the final section where some updating was necessary. When the time came for translating the entire book, both the translator and I felt that it would not help anyone to perform a fresh translation, ignoring the *Resonance* article altogether. Instead, we decided to use the article as much as possible, replacing the parts where the "conscious deviations" took place. In addition, some updating was necessary, as mentioned earlier.

I have longed to see an English translation of this book for a while. I approached several publishers, with negative results. I probably would have given up on the efforts but for the enthusiasm of my friend João Silva, who had read the *Resonance* article, heard the overall scheme of the Bengali book from me, and wanted to read the whole book in English. He actively participated in the efforts to find a publisher for the English edition. Finally, with his (and my own) efforts, we hit gold with a publisher as reputed as the publisher of this present edition. And then I hit gold again. I found that Dr. Sushan Konar was willing to undertake the task of translating the book. It is a matter of great fortune to have a scientist of her stature as the translator. In fact, she did more than translating: she went through the book carefully and initiated discussions with me, so that I began seeing my own book through her eyes. Sometimes this exercise resulted in minor changes in the text, as mentioned earlier in this preface.

The original Bengali book was published by the West Bengal State Book Board. I thank them for giving their kind permission for this translation. And finally, I want to take this opportunity to mention the name of my friend Syamal Das, who was instrumental in inspiring me to write the original book at a time when many of the ideas presented in the last two chapters of the book were novel even in the scientific community, and popular exposition of these ideas was not easily available, even in English. It was he who primarily dreamed the impossible dream. Like the Bengali edition, this translation is also dedicated to him.

Calcutta   Palash B. Pal

Translator's Foreword

A number of books have been written on the topic of the current book. "Why one more?" — this might be the question a reader would ask. The answer lies in the approach of this book, which will be evident for anyone taking even a cursory look at the contents. We are thankful to the team of CRC Press who realised the potential and the scope of this book far beyond that of its Bengali original.

But the question for a translator would be — "What is in it for me?" Actually, quite a lot. Working with Prof. Palash Baran Pal ("Palash-da" to his students and junior colleagues) has three advantages. First, the flow of his original text is such that the translation followed almost without any effort on the part of the translator. Then he took a very active part in the translation — going through the manuscript, correcting mistakes, making useful suggestions, and coming to the rescue whenever the translator got stuck. But the third and the final advantage is the most significant of all — it was an opportunity to learn (both physics and a lot of non-physics) from one of the greatest teachers and iconic figures of the Indian physics community today.

As a translator, it would probably be remiss not to mention here the people who made it possible for me to embrace English as an everyday language. But for Gita Choudhury,

Parimal Dasgupta, Pravrajika Jyotiprana, and Sudeshna Mukherjee I would still be one of the billions who have to go through life with a major handicap of not being comfortable with the most universal language.

Even though the work itself was extremely enjoyable, it would not have been possible without the support of my family. I am really grateful to Tirthankar for picking up much more than his fair share of parental duties despite his own busy schedule, and to our daughter Ele, who had to forgo a lot of her bedtime story sessions because her "Ma" insisted on keeping late hours!

Pune   Sushan Konar

Mathematical Symbols

ab : The product of two numbers, a and b.

a^2 : $a \times a$

$\sqrt{a}$: A number whose square is a, i.e., $\sqrt{a} \times \sqrt{a} = a$.

π : A special number, the ratio between the perimeter and the diameter of a circle. The value of this number is approximately 22/7 or 3.14159...

10^n : For a positive integer n this symbol means 1 followed by n zeros. For example, 10^2 is equal to 100, 10^6 is equal to one million and so on.

10^{-n} : Equal to $1/10^n$, i.e., one part in 10^n.

Chapter 1

Conservation laws

1.1 Preamble

Poets like to use fantastic similes. Essayists love great quotations. A singer's trump card is the staccato voice. A soccer player mesmerizes us with artistic dribbling.

Similarly we might ask — what do the physicists like?

Probably there is only one answer to this. Physicists like conservation laws.

What is a conservation law? Simple answer — this is a law which says that "some quantity is conserved". Then what does 'conserved' mean? It means that the quantity is not changing. Neither is it increasing nor decreasing, whatever its magnitude was seven seconds ago would remain so after two minutes or five years.

This is known as a *conservation law*. Then the question is — what are the quantities that are invariant, and what is their significance for the physicists and so on. The aim of this chapter is to follow the answers to such questions and paint a picture of classical physics.

1

Nobody can say how old the history of physics is, or exactly which discovery heralded the birth of this branch of studies. But the concept of conservation laws is not very old, perhaps 300 years at the most. In the days of ancient Greeks, i.e., the era of Ptolemy or Archimedes, a number of physical laws were discovered. But these did not include a single conservation law. Because Greeks were not concerned with the variation of a quantity with time. They calculated the pressure on the wall of a container exerted by water kept at rest inside it. But they did not say anything about flowing water, like that in a river. They told us how to calculate the minimum force required to lift a heavy object using a lever. But they did not say how much time such a process would take. The concept of time itself was beyond the purview of Greek science. Why? Perhaps they did not want to jump ahead of time! They wanted to look at issues that could be investigated without worrying about time. Perhaps they would have moved onto more complex issues later but before that their civilization itself was decimated by the Roman invasion. There could have been another reason. They did not have any good time-measuring tool. It really is not possible to do any serious scientific work with a sundial or an hourglass.

In the fourteenth and fifteenth centuries when Europeans started sailing to far corners of the world, they needed accurate time-keeping devices to keep the ships on their course. The technology for making better clocks advanced in the time that followed.

1.2   Conservation of momentum

Now began the work that remained ignored by the Greeks. Tycho Brahe collected detailed data regarding stellar and planetary positions in the sky at different times. These observations were organized into mathematical equations by Johannes Kepler. And wonderful experiments, using ordinary everyday objects, were performed by Galileo Galilei. The efforts of many such brilliant scientists culminated in Isaac Newton's time immemorial book Principia, published in 1687.

Newton's famous laws of motion are laid out right at the beginning of Principia. The first one says — unless acted upon by an external force, a material particle would either remain at rest or would continue to move at a constant speed along a straight line. The statement about the constancy of speed and motion along a straight line is usually summarized by saying that the velocity of the particle would not change. Both the magnitude and direction of velocity would remain constant. If the particle was at rest in the beginning, its velocity was zero. And it would remain zero for all time. On the other hand, if it was moving at three meters per minute in the northerly direction then it would continue to do so.

We almost have a conservation law in this. But before making it explicit, let us recall Newton's second law. This tells us what happens to a particle when force is applied to it. Newton said that the rate of change of the particle's momentum would be proportional to the applied force.

Let's not get puzzled by the word *momentum*. It is just the product of mass and velocity of a particle. If the word *mass* sounds unfamiliar, no problem. There would be an

explanation soon. In any case, it can be seen clearly that
the second law says — if zero force is exerted on a particle then the rate of change of its momentum would also
be zero. Meaning, the momentum would not change — it
would be conserved.

In the first law we stated that the velocity would not
change in absence of a force. Now we are saying that the
momentum would not change. There is little difference between the two — because Newton said that mass is simply
a measure of the amount of material inside an object. This
amount is not changing. Therefore velocity remaining constant is almost equivalent to momentum being constant.

Notice that we said 'almost'. Because there would be
some difference. When the number of objects is not one
but many we would see this difference. For example, consider a train with eighteen coaches, with a total mass of 900
tons. These were at rest on the railway tracks. Now an engine of mass equal to 100 tons needs to be coupled to this.
The engine reversed at a speed of ten kilometers per hour,
and then gave a push to the coaches. Those who have seen
the coupling of an engine to long-distance trains know that
after the push the engine and the coaches start moving together. The question is — how fast? Do they move at a
speed equal to the sum of their original speed? No. The answer lies in the fact that the total momentum of the engine
and the coaches remains the same after the push. The original velocity of the coaches was zero, hence their momentum
also was zero. The momentum of the engine was 100×10,
i.e., 1000 units. After the push, the mass of the entire train
became $100 + 900$, i.e., 1000 tons. Therefore its velocity became $1000/1000 = 1$ unit, meaning the train along with engine reversed with a velocity of 1 kilometer per hour.

Let us think about this a little. Have we bungled some-where? First we said that the momentum would remain constant. This means that no force is acting. But when the engine joined the coaches it gave those a push. Conversely, the coaches too must have applied an opposite force on the engine.

No, we have not bungled, this is correct. It is true that the engine is exerting a push, but we never said that the momentum of just the engine would not change. The coaches initially had zero momentum, afterward they started moving with a velocity of one kilometer per hour. Then why should we say that their momentum has not changed? We have not made this claim either. But no external force has been applied on the combined *system* of the engine and the coaches. And there could be any amount of push and pull between the engine and the coaches, any kind of processes could be taking place within the system — all of that would cancel each other out, according to the third law of Newton. Therefore, if there is no external force on the system then there is no net force on it — the total momentum would also remain constant then. Exactly that is what has been said here.

Of course, if we stop here it may suggest that the engine-coach combine would now keep going in the reverse direc-tion with a constant velocity forever — according to the first law of Newton. However, that does not happen in reality. As soon as the train starts moving, it encounters friction with the tracks, and with the air, then the driver of the en-gine applies brake, i.e., external forces come into play and the train stops within a little while because of that. And that's exactly what we observe.

Now we should think about this carefully. We have solved the problem very easily. If, in order to calculate the final velocity of the train after the engine was coupled to the coaches, we needed to know the magnitude of the force exerted by the engine, how this force was being opposed by the metallic frame of the coaches, and so on — it would have been an extraordinarily difficult problem. Instead we performed this calculation without even requiring a pencil and paper, simply using the idea of momentum conservation. No wonder the physicists would like such a beautiful concept.

We shall have to discuss other aspects of momentum conservation later. However, there is something that has to be mentioned right away. To give and example of velocity we had said — "three meters per minute due north". Notice the statement. It appears that a definition of velocity contains two facts — one is its magnitude (for example, three meters per minute), another is its direction (like, due north). Suppose the speedometer of a bus traveling from Kolkata to Kharagpur shows sixty kilometers per hour and the same happens in a bus traveling from Kharagpur to Kolkata — yet we cannot say that both the buses have the same velocity. They indeed have the same speed but their velocities are different. This is why velocity is called a *vector* — a physical quantity which requires both a magnitude and a direction to be completely specified.

If someone wants to know today's temperature, and if we answer 30° celsius, it satisfies the questioner. But if the question happens to be about the location of Mr. X and we say that he resides thirty miles far from here, the immediate query would be in which direction. This question arises because the position of an object is a vector. So is the velocity.

Similarly, the momentum too is a vector. This means that the conservation of momentum would require both its magnitude and its direction to be conserved. If more than one object is involved, we need to remember this question of direction. Otherwise our calculation would not give us the correct answer. If the momentum of an object is five units due east and it is three units due west for another then the total combined momentum of these two objects would be two units due east. Force also is a vector. In a game of tug of war, if one team pulls the rope towards north with great force and the other team pulls it towards south with equal force then the two forces cancel each other. And we find the rope not moving in any direction at all. That is what should be expected in the absence of any force. This means that if we take the north as the direction of positive force, the south would be considered as the direction of negative force and vice versa.

If a person stays in a northeasterly direction, what would happen then? Then we would perhaps say — first go five kilometers due north and then move five kilometers due east. And then if we add that this person lives on the third floor — we have really completed our job. We have now resolved this person's position vector into three parts — one along the north-south line, another along the east-west line, and the third along the vertical. Similarly, velocity, momentum or force could also be resolved into these three parts. The three numbers obtained this way would give us both the magnitude and the direction of the vector. In technical language these three parts of a given vector are called its *components*. The conservation of momentum actually means the conservation of its components separately. That is, even though the conservation of momentum looks like one law, it actually contains three conservation laws within it.

1.3   Conservation of angular momentum

But momentum conservation was not enough for Newton. Let us see now why that was so.

Kepler had explained how the Earth or other planets rotate around the Sun. But his solutions were exclusively for the Sun and the planets. He did not say anything about any wider application of this concept to other situations. The challenge before Newton was to explain the motion of heavenly objects by the same laws of force describing the motion of objects in our ordinary everyday life.

Let us consider the Earth. It is going around the Sun — this means that the motion is not along a straight line. The direction of its velocity is continually changing, whatever might be its magnitude. This means some external force must be acting on it. Of course, that is true. The Sun is attracting the Earth. Newton said — there is nothing special about this between the Earth and the Sun. In reality, every object in this universe is exerting an attractive force on other objects. As soon as we say that there is an 'attractive force' we know the direction of that force. And what would be its magnitude? Newton's answer was this — if the masses of the two objects are m_1 and m_2 and the distance between them is r then the magnitude of this force is given by —

$$\frac{Gm_1m_2}{r^2}.$$

This is Newton's famous *law of gravitation*. Here G is a universal constant. This means that whatever may the two objects or their intervening distance be — to obtain the gravitational force we would always have to have the same number as a pre-factor.

In any case, one thing is certain. A force is being exerted on Earth since its momentum is not conserved. So our newly learned idea of momentum conservation is quite useless here. What then?

No need to despair. We know the force and can therefore use Newton's second law of motion. This would give us the rate of change of momentum. If we know the momentum at a given point of time, we can find the momentum in the next instant — since the rate of change of momentum is known. From there onto the next instant and so on. We can proceed in this manner. And momentum divided by mass is the velocity. Then we would have complete knowledge about the orbit of Earth.

Newton produced the solution in this way. Even now the movements of planets and satellites are calculated this way. But is it possible to save on the effort using a conservation law? Do we have any hint from Kepler's laws?

Yes, indeed. Firstly, Kepler says that each planet moves in an elliptical orbit around the Sun, with the Sun in one of the focal points of the ellipse. An ellipse is like a flattened circle, somewhat like an egg. And it is not really required to know the exact definition of the focal point of an ellipse, it is enough to look at Fig. 1.1. Here S denotes the Sun.

In any case, since the orbit is not a circle the distance between a planet and the Sun is not equal at all times. When the planet is nearer — say at point A, then due to a smaller distance the force exerted by the Sun would be

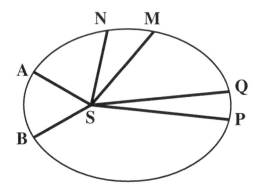

Figure 1.1: Earth orbiting around the Sun according to Kepler's law.

larger according to Newton's law of gravitation. Therefore, the planet is given a larger momentum. Now say in four weeks time the planet goes from point A to point B. On the other hand, when the planet is at a point P, the distance is larger and the gravitational attraction is smaller. Then the planet would perhaps move from point P to point Q in four weeks. So the movement is through a much smaller distance. But if we measure the area of the regions ABS and PQS then we would find them to be equal. If the planet moves, from M to N during another four-week time span, then the area 'swept out' by the planet is MNS. And this too would be equal to ABS or PQS. This means that during any four-week time period the planet always traverses through the same area.

This definitely is a conservation law. Now the task is to apply this concept to a larger context, instead of confining it only to the movement of planets around the Sun. In the end, this gave rise to a law very similar to Newton's second law of motion — the change of angular momentum of any object is equal to the torque applied to it. Now we can easily say that if there is no external torque applied to a system, its angular momentum would be conserved.

So we have another conservation law, but we do not have a clear idea about it yet — because we still do not know what is meant by torque or for that matter angular momentum. Well, let us get down to that then.

Consider a wheel. Or a compact disc. The disc is not rotating at the moment but is stationary. The centre of this disc is at point O in Fig. 1.2. We want to rotate the disc by applying a force at the point P. We have not yet decided in which direction to apply the force.

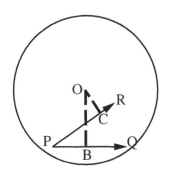

Figure 1.2: Torque required to rotate a disc.

Let us say we apply this force along PQ. The disc now starts to rotate. But if we applied the same force along PR, the disc would not have rotated as fast. This means that the speed of rotation does not depend on the magnitude of the force only. The mystery lies deeper.

Let us draw a line from O perpendicular to PQ. If we multiply the length of this line OB to the magnitude of the force applied along PQ, then the quantity obtained is the *torque* applied, with respect to the point O. If the force is applied along PR then the perpendicular line has to dropped onto PR. In Fig. 1.2 we can see that the length of this perpendicular, i.e., the line OC, is half of the length of OB. Therefore the magnitude of the applied torque in this case would be half, in spite of the force being the same. However, if the force applied along PR would

Figure 1.3: A direction of the spin acquired by a cricket ball depends on the direction of the torque applied.

have been double, the torque in that case would have been the same and the disc would have rotated at an equal speed. It is clear now that the speed of rotation depends on the magnitude of the torque and we have also learned how to calculate it.

Now let us look at Fig. 1.3. A spin bowler is coming to bowl. The way he holds the ball can be seen in here. Just before releasing the ball he would twist his wrist, meaning the cricket ball would get a torque. But in which direction? If the wrist turns in the direction of the arrow drawn in the picture in the left the batsman would see an off-spin ball. If the wrist turns according to the arrow in the middle picture, it would be leg-spin. And if the bowler applies the torque by pulling his index finger backwards, as seen in the picture in the right, then it would be backspin bowling. This means that not only the magnitude, the direction is also important for torque. In case of off-spin and leg-spin we would say that the torque is perpendicular to the plane of the paper. For off-spin, the arrow is rotating anti-clockwise. The torque would be upwards through the paper. In case of leg-spin the arrow is rotating clockwise and the torque would

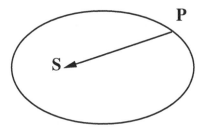

Figure 1.4: Force on a particle moving in an elliptical orbit.

be downwards through the plane of the paper. The torque would be towards the right of the paper in case of backspin, because from that direction the arrow would appear to rotate anti-clockwise.

This means that the torque is also a vector. We need to know both the magnitude and the direction of this. Unless we know the direction there can be no difference between an off-spin and a leg-spin. What a disaster that would be for a batsman!

Similarly, if we draw the momentum of P in Fig. 1.2, multiply its magnitude by the perpendicular distance form point O, then we obtain the magnitude of the angular momentum of point P. The direction of this would also be obtained by the same rule. Now we understand what torque and angular momentum are. And the law says — the rate of change of angular momentum is equal to the applied torque.

Let us then consider the situation with the planets. We can decide the point with respect to which the torque would be calculated. Let us find it with respect to the Sun. (Or with respect to the center of the Sun, for a more critical reader.) We need to look at Fig. 1.4 now. The Sun is at the point S, and the planet is at P. We have learnt about the law of grav-

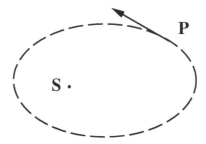

Figure 1.5: Momentum of a particle at a given point of its elliptical orbit.

itation already. The Sun is attracting the planet along PS. If we try to draw a perpendicular on this line from S, then the length of that perpendicular would be zero. Because S itself lies on this line. Since the perpendicular distance is zero, the torque is also zero. There is no torque.

This means that the angular momentum of the planet would be conserved. If the area traversed by the planet in its orbit, mentioned in the context of Kepler's law, is multiplied by $2m$ (m is the mass of the planet) then we obtain the angular momentum of the planet. According to Newton's definition, m does not increase or decrease. Therefore, Kepler's law and the conservation of angular momentum are actually the same thing, the last one being more general with far wider applications.

We have given an example of how convenient it is to use the conservation of angular momentum. Let us quickly look at a similar example here. We have drawn Fig. 1.5 just like Fig. 1.4. In Fig. 1.4 we had shown the force applied to P, now in Fig. 1.5 we show the momentum of P — this is the only difference. The picture is like that of the off-spin ball, that is, the angular momentum is perpendicular to the plane of the paper, in upward direction. This is the state at a particular instant.

But we already know that this angular momentum is conserved. Therefore the direction would not undergo any change, it would always be perpendicular to the plane of the paper. That would imply that the planet would always have to move on the plane of the paper, it can never leave this plane.

Note that this is not a small issue. Without doing any calculations we have found that the planetary orbits around the Sun always lie in a plane. Once again this shows the worth of conservation laws.

Therefore, Newtonian mechanics has given us these two conservation laws — of momentum and angular momentum. Since both of these have three components each, it is more accurate to say six conservation laws in total. Of course, these cannot be applied to any arbitrary system, we need to make sure that there is no external force or external torque applied to the system.

However, if our system happens to be the entire universe we are in an excellent situation. There is no question of any "external" force or torque, because there is no outside at all. The total momentum of the universe is therefore conserved and the same is true of its angular momentum.

Newtonian mechanics also provides for another conservation law. We are talking about the energy conservation. But history tells us that it appeared almost 150 years after Principia. We need to wait for that. Meanwhile we can look around at other branches of physics.

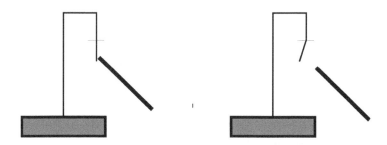

Figure 1.6: Franklin's experiment demonstrating the attraction between unlike charges and repulsion between similar ones.

1.4   Conservation of electric charge

If a piece of resin is rubbed by wool, it is seen that the resin starts attracting very light objects. This phenomenon has been known for thousands of years. Earlier, people used to think that it is a special property of resin.

William Gilbert demonstrated at the end of sixteenth century that many other materials like glass, sulfur or lac, besides resin, also have this property. Gilbert said that rubbing enables these material with an ability to to apply a special kind of force. He named this force as the 'electric' force.

Of course, detailed experimentation with electricity started much later, in the beginning of eighteenth century. The superintendent of the "Jardin du Roi" (literally meaning "Garden of the king" or the "Royal Gardens") of the French emperor was a gentleman called Charles du Fay. He proposed a theory in 1733. He rubbed a glass rod with silk. The rod became charged. This means that the rod now had the power to apply electric force. On the other side a thin gold leaf was hanging from a peg, which was mounted on a block of wax. The arrangement is shown in Fig. 1.6. He touched the gold leaf with the glass rod and then took the rod away. Then he brought back the glass rod near the gold leaf and saw that they were repelling each other now.

It was understood that electric forces attract. But now we find a case of repulsion. du Fay reflected that when the glass rod touched the gold leaf, the leaf became charged. Thereafter charged rod and charged leaf repelled each other. Does this mean that two charged objects repel each other, while an charged object attracts a non-charged object?

But more surprises awaited. du Fay observed that if instead of the glass rod, a piece of resin, rubbed by wool, is now brought near the gold leaf, then they attract each other. This means that sometimes there could also be attraction between two charged objects! From this du Fay concluded that there are two kinds of electricity. One kind of electricity is generated when silk is rubbed against glass and another kind is generated when wool is rubbed against resin. Just like water flows down when a container filled with water is tilted, electricity has flowed like a liquid to the gold leaf when the glass rod touched it. Thereafter it has been experiencing repulsion to similar kind of electricity and attraction to the other other kind. du Fay explained the results of his experiment by these two types of *electric liquid*.

In 1745, Benjamin Franklin of United States of America had traveled to Europe. He was fascinated by a machine there. On one side of it someone would turn a handle. The way du Fay had rubbed silk onto resin, the turning of this handle would rub some material against another. Then electricity would accumulate in the material placed at the other end of the machine. After returning to America he imported this machine from Europe and started doing his own scientific experiments.

Franklin conducted a lot of experiments. One of these was like this. Two persons, A and B, are standing on a block of wax, along with the machine brought by Franklin. A is turning the handle of the machine and B is touching the machine on the other side. A third person, say C, is standing on the floor. If B touches C after A has turned the handle for a while, C would immediately experience a shock. If the shock is really strong even sparks would fly. This means that B has been charged by the turning of the handle. On the other hand, if A touches C, the same thing happens. Meaning, not only B, A too has been charged. Moreover, Franklin saw that if A touched B after turning the handle for an equal amount of time then the shock is much stronger, the spark far brighter. If A or B touches C after that then no shock is experienced.

If someone is asked for an explanation of this today, it can be given very easily. We all know that every object is made up of atoms. A positively charged nucleus resides at the centre of an atom. And a cloud of negatively charged electrons are orbiting around this nucleus. Initially, the amount of positive and negative charges are equal in each person, making all of them charge neutral, i.e., they all have zero charge. Now A turns the handle. As a result some of the electrons travel from his body to B. Now B has some extra amount of negative charge — say two units. But this charge has come from A, therefore he has a deficit of negative charge. So A has a relative excess of positive charges. If we measure this we would find that the amount of positive charge in A's body is exactly two units again. They are standing on a block of wax, which is a very bad conductor, and therefore the charge cannot flow out of their bodies. Whereas C is standing on the floor. That is why if B now touches C, the charge would flow to the floor through C's body and he would get a shock. Whereas if A touches C, some electrons would flow from the floor to A through C's body. Again C would experience a shock. But there exists a difference of four units of charge between A and B. So if they touch each other directly, the extra electrons of B would neutralize A's charge with double the fervour.

There are two main differences between this explanation and that of du Fay. du Fay said that two types of electric liquid are being produced and the flow of such liquids from one object to another is giving rise to all the observed phenomena. Here we see that only one type of particles (electrons) are moving. Of course, the positively charged nuclei are also there. But they are a few thousand times heavier than the electrons and consequently are quite sluggish. Therefore, we have only 'one type of particles' not 'two types of liquids' at the root of all electric phenomena — an excess of this particle gives rise to negative charges and a deficit to positive charges.

Secondly, whether we call this an electric liquid in the old style or electrons according to the modern jargon — these are not being 'created' by rubbing. Rubbing glass with silk, du Fay found it to be negatively charged. If he had looked at the piece of silk, he would have found a net positive charge, meaning some of the electrons from silk had moved to glass. No charge had been created or destroyed. The amount of charge had remained the same — it had only been redistributed, whereby some objects displayed an excess and some deficit.

In other words, the total amount of charge is conserved. This is another of our conservation laws, first stated by Benjamin Franklin in 1747. Franklin did not mention electrons or nuclei because they would be discovered another 150 years later. However, except for that there is no difference between his explanation and the modern understanding of it.

1.5   The beginning of chemistry

We shall have to go out of the purview of physics now and see what the European chemists were doing in the post-Renaissance period.

In the middle ages, chemistry was another name for magic, a bag of tricks to fool people. Renaissance liberated chemistry from the shackles of such lowliness. In the first flush of freedom the process that was investigated in detail by the chemists was combustion.

When wood is burnt, ashes remain and the smoke goes away. Why? A group of scientists said that everything contain a kind of particles called 'phlogiston'. When an object is burnt the phlogistons are released. What remains is the ash.

In ordinary everyday usage combustion means burning by fire. But it has a more general meaning to the chemists. The reaction of any material with the air is known as combustion. In case of wood, the process starts at a rather high temperature and typically fire is needed for the job. On the other hand, if a piece of iron is left alone, it slowly reacts with the air to produce rust — that is a kind of combustion too.

If we accept the phlogiston theory then we would expect the weight of the piece of iron to decrease. Quite contrary to this expectation, the weight of iron increases when a layer of rust covers it. Not only iron, the weight of almost all metals increase as a result of combustion. By the mid-seventeenth century a number of scientists confirmed this fact by their detailed experimentation.

The phlogiston theory did not really get destroyed by this completely. In fact, there were lot of efforts to explain the observations by suitably modifying this theory.

But there was a problem. Whether it was actual burning or any other kind of chemical reaction, the released gas escaped from the system and mixed with the ambient air. If instead the reaction takes place inside a closed container then nothing can escape. Any element required for the reaction can come from the air within the container itself. And if a gas is produced in the reaction that also stays put inside.

Till then, wood was being burnt in open air. Most of the end products of this reaction are gaseous. The left over mass is very small. If now the wood is burnt inside a container, then the total mass of the air, wood and container should be the same before and after the process of burning. If a piece of iron is left outside, the oxygen from air would combine with iron to produce rust and as a result the mass of iron would increase. If iron rusts within a closed container the total mass would not change.

The total mass always remains conserved in a chemical reaction. In all likelihood, Lomonosov from Russia stated this for the first time in 1748. But Russia was disconnected from the main stream of research in Europe and nobody really knew about his discovery. Thirty-seven long years after this, Lavoisier in France discovered the same law after a lot of experimentation and immediately this message reached the scientists of England, France and Germany.

Then what exactly is happening in a chemical reaction? Even if the mass is not changing, something else must be changing. What is that? That question became important now.

The answer to this came in 1803, from John Dalton. Dalton said everything is made up of atoms. All the atoms are identical in certain materials, they are called *elements*. This is what we see in metals like iron, copper or lead. These metals are therefore elements. Similarly, oxygen or nitrogen found in air are also elements, because here too all the atoms are identical. When a piece of iron is left in air, the atoms of iron and oxygen combine to make a *compound* or the molecules of a compound. This means that all the atoms of a compound are not identical.

So we have the answer to one of our questions. What changes in a chemical reaction is the combination of different kinds of atoms to form molecules. Tweedledee's sack on Tweedledum and Tweedledum's sack onto Tweedledee!

But.... Dalton said, it needs to be remembered that the total number of atoms remains conserved. Hydrogen and oxygen combine to form water. This means that two atoms of hydrogen and one atom of oxygen combine to make one atom of water. This is why the chemical symbol for water is H_2O. But the number of atoms is again three, two hydrogen (H) atoms and one oxygen (O) atom.

So the underlying truth about chemical reactions becomes clear by the use of a conservation law.

1.6   Conservation of energy

The machines built in the first phase of industrial revolution were definitely much more powerful and efficient than men. Many of the manual tasks, like running a loom or copying a book, were now being done by the machines. But carrying load was still being done by horse-drawn carriages. Because more efficient transport solutions were not yet available.

But the new capitalist class was not happy about this. Mechanized vehicles could carry the factory produced clothes to faraway places. This would mean more sales and consequently more profit. Also the power-loom consumed too much power. The effort was to get more work out of the same amount of fuel.

Therefore, they started building research institutes, arranged for various competitions — to find who can build the fastest engine, who can find a way to mine coal in the least expensive way and so on.

Soon, the results of such research started coming in. The machines became more efficient. Along with it increased the greed of the capitalists. More, and even more. In the end, it appeared that almost the whole of Europe had become obsessed with a peculiar idea. Couldn't one make a machine which once started would go on forever? In other words, is it possible to make a perpetual motion machine?

We shall not discuss that history here. Suffice it to say that even after a lot of effort nobody was able to invent such a machine. And the scientists started thinking why that was so? Why would a perpetual motion machine not exist?

Slowly the answer became clear by the effort of many researchers. The most important steps came from Hermann Helmholtz, in 1847.

Let us assume that we have a number of objects. The system comprising these do not have any external energy being given to it. Of course, they have their own mutual interactions. Helmholtz showed that according to Newtonian mechanics the sum of the kinetic and potential energy of those objects would be conserved.

What is kinetic energy? The energy contained within an object due to its motion is its kinetic energy. When we throw a stone at a mango tree, the kinetic energy of the stone detaches the mango from its stalk making it fall. The larger the velocity of the stone, the more effective it would be in detaching a mango from a firmer stalk, because the stone would have larger kinetic energy.

There could be other forms of energy besides the kinetic. When a book is kept on a shelf, we know that even though it does not have any kinetic energy it contains some energy by virtue of its position. If the book falls from a height it can break a glass. Meaning, the book has *potential energy*. The higher the book is stored on a shelf, the larger potential energy it would have.

Accurately speaking, when the book falls, it loses height continuously and consequently its potential energy decreases. On the other hand, its speed increases, increasing the kinetic energy in the process. So kinetic energy is increasing here at the expense of potential energy — the sum of these two remaining constant. Finally when the book falls on the glass, the kinetic energy breaks the glass. And the book lies on the floor losing all its potential energy.

And what happens when the book falls to the floor directly? There would be a sound like a loud clap, when the potential energy is finally transformed into sound. The sound is a type of energy too. This means that in this instance also, the energy is just changing its form. Not only sound, there are many other forms of energy — like heat, electricity, or magnetic energy. And we see lots of cases of energy transformation, from one type to another, in our everyday life. When a switch is pressed, electrical energy flows through the wire and then gets converted to heat and light in the filament of a bulb. Burning coal generates heat from which a steam engine gets its kinetic energy. In modern engines this energy does not come from heat but from electricity instead. When we clap, the kinetic energy of the hands gets converted to sound. And rubbing hands on a cold winter morning gives us heat at the expense of kinetic energy.

But where does this heat go? It energises the air molecules a little. Where does the sound of the claps go? The air molecules vibrate a little, and gets heated a bit. Though these changes are so small that we cannot discern them and we may think that the energy gets completely destroyed.

That surely does not happen. Energy can only transform, there is no creation or annihilation of energy. If a system does not receive energy from outside and there is no flow of energy from the system to the outside, the total energy of the system would be conserved. The total energy of the whole universe is, therefore, conserved. Because there is nothing 'outside' the universe and so energy cannot flow in from or flow out to anywhere. No machine can therefore go on forever, because it can only generate as much energy as it has been given to begin with.

This is what is called the *law of energy conservation*.

1.7   The danger

We have familiarized ourselves with all the conservation laws that classical physics gifted to us. But we were so pre-occupied with picking up these conservation laws one after another that we did not keep track of our collection properly. Our story has now reached the middle of the nineteenth century. Do the ancient conservation laws of momentum or angular momentum still hold true or have they been upturned by some new discoveries?

To be honest, it was not a good time for these. The problem had been created by a brand new branch of physics — current electricity. This branch was developed in the beginning of the nineteenth century, when Volta made the first electric battery and sent the first electric current. Thereafter it was enriched by the work of scientists like Ampere, Oersted, Ohm, Faraday, Henry etc.

Something peculiar had been observed. If current flowed through two wires then the force applied by the second wire on the first would be along the direction of the perpendicular to

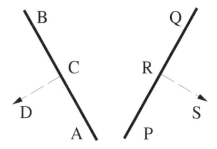

Figure 1.7:   Electromagnetic force between two current carrying wires.

the first wire. Similarly, the first wire would also exert a force on the second wire along a perpendicular onto it.

We need to look at Fig. 1.7 to understand why this is peculiar. AB and PQ are the two wires. When there is no current, they are stationary, and the total momentum is zero. Now current is sent through the wires, which made the wires exert force on each other. But there is no force on the system from outside, hence the momentum should still remain zero. That is what we have learnt.

But we see that the forces are acting along the dashed lines CD and RS. So the movement would be along these directions — the momentum of AB and PQ wires are therefore along CD and RS respectively.

If the momenta of the two wires were in opposite directions and if their magnitude were also equal, then their sum would have been zero. We would have been happy with our calculations. But here even the directions are not opposite to each other. There is no question of the sum being zero, whatever be their magnitude.

This means that the momentum is not the same as what it was in the beginning. Under the circumstances there are only two options. One option is to say farewell to momentum conservation now. The other is to do some soul-searching, i.e., to ask whether we were right in assuming that there is no external force acting on the system, i.e., there is no flow of momentum from or into the system comrpised of two wires.

We have said in the beginning that physicists love conservation laws very much. So they are extremely reluctant to take the first option. But the situation is really complicated. Not only momentum, we have the same problem with angular momentum. Even the situation with the energy conservation is in trouble.

We have not worried about one question though. Let us say that a candle is burning. We have been asked to extinguish it. This can be done in either of the two ways. One: we can hold the wick with a big forceps. Two: we can blow on the candle. The air from the blow would flow to the wick of the candle, which would extinguish the fire. So the first method uses direct contact to extinguish the candle. But the second? Let us say that a guest from planet Mars is watching this and he does not know that when we blow air gushes out from our mouth. He may describe the process like this — 'someone bulged his cheeks at a distance and the candle got extinguished here'. But that is not what actually happens. A gust of air has been sent to the candle which has extinguished the fire. It does not make any sense to talk about the incident without mentioning this apparently invisible connection between the candle and the mouth of the blower.

Now let us see what happens in the case of the two wires. The wires are not touching each other. Then how could the first wire feel the force exerted by the second wire? Which invisible link binds them now? No, it is not air, neither is it any other particles. Because if the system is completely evacuated by a pump even then the situation in Fig. 1.7 would not change at all.

The answer that came out through the work of Faraday, Maxwell, etc., is like this. Because of the current flowing through a wire, an *electromagnetic field* is set up around the wire. Such a field has been created in an area around the wire AB in Fig. 1.7, which encompasses the wire PQ. Therefore the field is exerting a force on PQ.

Let us understand this. The actual reason for the extinguishing of the candle is not bulging of the cheeks. Bulging of the cheeks simply produces a gust of air which has extinguished the candle. Similarly, AB has not exerted a force directly on the wire PQ. AB has generated a force field and this field has exerted force on PQ. On the other hand, PQ has also generated a force field which encompasses AB. That is why AB too experiences a force.

Earlier we thought that the system comprising the wires AB and PQ have no connection with the outside world. Now we see that such is not the case. At least, we need to take into account the electromagnetic field. Otherwise there would be gaps in the explanation.

Maxwell said, that is exactly what has happened. We did not notice that the field itself may have momentum, angular momentum and even energy. If we exclude these from our calculations, it would not be correct.

But momentum in this case would not be obtained by multiplying mass to the velocity. Because there is no way to define the mass of a field. Maxwell and his followers have taught us how to calculate the momentum of a field. The angular momentum and energy can also be obtained. And after these quantities were taken into account, it was found that all the conservation laws were now being obeyed perfectly.

Now the readers would ask: how does the Earth then feel the gravitational attraction of the Sun, given that it is certainly not touching the Sun? Ah, yes. There too we have the gravitational field, exactly like the electromagnetic field. Newton did not think about it. Einstein did, more than 200 years later. His theory of general relativity discussed the gravitational field. This gravitational field also has momentum, has angular momentum. But the magnitudes of these quantities are very small. Consequently, we did not face any problem ignoring these. However, if instead of the Sun we had a very massive star in our backyard, then the gravitational attraction would have been very strong. And we could not have done without taking it into account. We would have had similar problems to explain the movements of planets, like we have observed in the situation described in Fig. 1.7.

1.8   The synthesis

Maxwell had understood another important issue. When we discussed various forms of energy in Sec.1.6, we had separately mentioned electric energy and magnetic energy. While constructing his theory, Maxwell realised that it was not possible to separate these two. So he talked about the electromagnetic field, not the electric field or the magnetic field separately.

Not only that. The light waves too are electromagnetic waves in reality, except for the fact that their wavelengths are just right to stimulate the human eye. From the point of view of physics, there is no difference between the light-wave and the electromagnetic wave we tune our radio sets to. It does not make much sense to talk about the light energy, the electric energy or the magnetic energy separately. All of these are actually the energy of the electromagnetic field.

The dream of any physicist is unification. The final aim of physics is to explain everything in the universe by one single theory. For the first time, Newton gave us a taste of such unification by explaining the motion of planets and the motion of everyday objects using the same laws. Maxwell placed another brick on the wall by explaining the nature of light, electricity and magnetism by his theory of electromagnetism.

Another aspect was also becoming clear through the efforts of scientists like Maxwell, Boltzmann, Meyer, Gibbs, etc. The atmosphere exerts pressure on the Earth. The gas inside a container exerts pressure on the walls of the container. Why? These scientists said that the molecules of the gas are continuously colliding with the walls of the container. The force exerted by the collision of a single gas molecule is very small. But the number of gas molecules is so large that the effective force is not insignificant. This force is the reason behind the pressure.

So the gas molecules are in constant motion. But this is not organized motion. If all the molecules could have moved in a synchronized fashion, they could have pushed the entire container. Like a strong wind can blow a leaf away. But that is not the case here, the motion is completely random. The total momentum, therefore, is zero.

But if the molecules have motion, they also have kinetic energy. And kinetic energy is not a vector which could cancel out to zero. The scientists then realised that the kinetic energy due to this random motion is actually the heat energy. When they assumed this, all the results of their calculations turned out to be quite satisfactory.

What would happen if the gas is now heated? The molecules would start moving faster. The motion is still random. Hence nothing would happen to the momentum, but the total kinetic energy of this random motion would increase. We shall say that the gas has acquired some heat energy.

Not only gases, the scientists started believing that the same thing would happen even in liquids and solids. The molecules are extremely close to each other in a liquid or a solid, and therefore it is very complicated to calculate the force of inter-molecular attraction or repulsion. Yet the belief came into existence in the nineteenth century itself, though the actual calculations have been carried out in the twentieth century.

Therefore, heat energy is nothing but the kinetic energy. If the motion of the particles is organized, all particles move in the same direction, and the system has a net momentum — the energy is known as the kinetic energy. If however, the motion is random, the momenta of the molecules cancel each other to produce a net zero momentum then the total kinetic energy of the molecules is known as heat energy. This is the only difference.

By now, we also know that sound too is a form of kinetic energy. We see the pendulum oscillate in an old fashioned grandfather clock. The pendulum oscillates once in every second. If instead the pendulum had oscillated between 20 and 20,000 times per second, the energy of that vibration (that is the kinetic energy of the motion) induced on the neighbouring air molecules would have been able to stimulate human auditory senses. Then we would have said that we are hearing a sound.

We have progressed even more along the path of unification now. We have very coarse vision, that is the only regret. If we had finer vision, right from the beginning we could have said that material particles can have only two types of energy — kinetic energy and potential energy. And yes, besides the particles there exist the electromagnetic and the gravitational field. These fields too can have energy. But that's it. The list ends here.

If we keep the consideration of fields aside for a moment, then the law of conservation of energy simply means that the total sum of the kinetic and the potential energy of the particles is invariant. This statement made by Helmholtz in 1847 actually included almost all kinds of energy.

We said 'almost' because now we need to add the energy of the fields. And then we would have a perfect result.

1.9 Relativity

By now, the nineteenth century was coming to a close and the twentieth century begun. This was a rather eventful century, right from the beginning. The quantum theory came into existence in 1900 itself — we shall discuss that in the next chapter. In 1905, Einstein proposed the theory of relativity. The main points of this would be outside the scope of our discussion here. We shall only refer to some of the results.

It was seen from the theory of relativity that the conservation of momentum and the conservation of energy are not two different concepts but are different ways of looking at the same thing. Even though we can definitely smell another unifying notion, there is no need to worry about that here. In fact, we can completely forget about it. Instead, let us see what else we have learnt from the theory of relativity.

Consider, for example, two colliding particles. There is no external force. We shall expect the momentum to be conserved. We know how to calculate the momentum of a material particle. Momentum is a vector, its direction is the same as that of the velocity, and its magnitude is equal to the product of the magnitude of the velocity with mass. Einstein showed that a vector is indeed being conserved and its direction is also along the direction of the velocity. But if the magnitude of the velocity is v then the magnitude of this conserved vector p is given by,

$$p = \frac{mv}{\sqrt{1 - (v^2/c^2)}}.$$

So, if we still want to define momentum as the product of mass and velocity we shall have to say that the mass is now given by,

$$M = \frac{m}{\sqrt{1 - (v^2/c^2)}}.$$

The mass of a material particle is fixed but it increases with the velocity. Meaning, the mass depends on the velocity. This is why sometimes it is referred to as the *kinetic mass*, which is denoted by M. And what is m then? It is the *rest mass*, because this is the mass of the particle when its velocity is zero.

The mass is dependent on velocity then. But Newton stated that the mass of an object remains invariant, since the mass is a measure of the material contained in the object. Einstein said that this definition is very confusing. What could be the meaning of the 'measure of material contained'?

There is no point in worrying about such inaccurate definitions. Instead let us make one thing clear. If the mass of an object is large, more effort is needed to move it. It requires more effort to lift a filled box than an empty box. Because the mass of the filled box is larger. Thus, a more operational definition of mass would be the resistance an object offers to efforts of moving it, or the inertia of the object. If we look at things this way it becomes clear that the mass increases with velocity, exactly the way Einstein has said it would. It has been proved in many experiments after the advent of the theory of relativity.

How could this be? When a planet orbits around the Sun, its velocity changes continuously. According to Einstein its mass would also change. We never considered this, but our calculations have completely matched the observations. We never had any problems applying Newton's laws to the motion of trains, of pendulums, of rockets in space and so on!

We did not run into trouble because the formula describing the kinetic mass has a quantity called c in it. This c is the velocity of light in empty space, which is 3×10^5 kilometers per second. No, not per hour or minute but per second! And what are the typical speeds of objects that we usually see? A fast train can move at about 100 kilometers per hour, that is about 30 meters per second — not even ten millionth of the speed of light. An aeroplane perhaps goes at 200 to 300 hundred meters per second. Even a rocket would move at about fifty kilometers per second. The speed at which the Earth orbits around the Sun is also of this order. All of these are simply extremely tiny compared to the speed of light. Because of this the ratio v/c becomes so small that the difference between m and M becomes totally irrelevant. This is why we cannot observe the change in the mass with speed. In modern accelerators, like cyclotron, betatron etc., particles like electron or proton are accelerated to speeds close to c. Then the increase in their mass is clearly observed. A single electron can become heavier than a loaded truck if its velocity can be appropriately increased. Therefore, when we still say that the mass of some object is smaller than something else, we usually refer to the rest mass of those objects.

Then again, Einstein showed that the energy of an object is related to this kinetic mass through a very simple relation given by,

$$E = \frac{mc^2}{\sqrt{1-(v^2/c^2)}} = Mc^2 .$$

So if an object stays at rest, it would still have some energy. The magnitude of this energy is mc^2 and this is called its *rest energy*. Not the potential energy but it is the rest energy. No need to keep the object on a high shelf for this. Wherever the object may be, it would have this amount of energy. We can say that this is the internal energy of an object.

Now if we impart a velocity to the object, the quantity $(1-v^2/c^2)$ would decrease and E would increase. This excess energy is the kinetic energy of the object. This can be written as follows —

$$\text{kinetic energy} = Mc^2 - mc^2 = mc^2 \left(\frac{1}{\sqrt{1-(v^2/c^2)}} - 1 \right) .$$

Here too, if the quantity v/c happens to be small we get back Newtonian laws. If that is not the case then we have to do our calculations using these formulae.

While discussing the transformation of energy, we need to consider the potential energy as well. If not we would be in trouble. How do we get so much heat by burning coal, otherwise? We require an initial bit of energy to make the coal catch fire but the total amount of heat obtained from coal is much larger than that. We may go back thinking that the energy conservation is not being obeyed here. That is not true. In reality, a part of the rest mass has been converted into heat. The ash and the gases obtained after burning would have a rest mass slightly smaller than the mass of the coal. Of course, the amount of heat is so small in ordinary chemical reactions that the change in the rest mass would be very difficult to observe. Still there would be discrepancy and it would be possible to measure that using accurate instruments.

Therefore, rest mass is not conserved, it is the kinetic mass that is conserved. We talked about conserving mass by making the reactions happen inside a closed container. Now we need a container that would not only confine the material, it would not even allow any energy to flow in or out. Because even a flow of heat would mean an exchange of energy and therefore exchange of mass with the outside world. And that would put a spanner in our calculations.

In the end then we come to the conclusion that mass is nothing but another name of energy, we just need to divide energy by c^2. If energy is conserved, then a quantity obtained by dividing the energy by two or seven would also be conserved. There is no doubt about it. Similarly, if the energy is divided by c^2, that too would be conserved. Therefore, there is no need to separately talk about the conservation of mass, or the conservation of kinetic mass.

1.10   Noether's theorem

To be precise, all we have discussed are just a few examples. We still have not discussed the most important fact about conservation laws in general. We shall try to understand that in this section.

A physicist performs his experiments in the laboratory, observes various events and tries to formulate general laws on the basis of those. He is pleased if the laws produce correct results. If not he discards his propositions unceremoniously. And tries again.

It is possible to challenge him about this. We could say that the observations he makes in his laboratory may not be the same as what happens in a faraway place. We might argue that things may happen differently for a different observer.

The physicist would answer that there is absolutely no chance of that. The laws of nature are not like the laws governing a country which change the moment one crosses the border. Laws of nature are applicable everywhere, they are invariant.

Of course, in a different environment the results of an experiment would be different. But if we contrive to set up identical experiments, then the results would also be identical wherever they may be conducted. In fact, if some magical monster were to move our universe by some enormous distance one day, we would not have noticed its effect at all. All the everyday activities would continue to happen as usual.

In scientific terminology we call this the *homogeneity of space*. Homogeneity means that there is no special point in space. Every point is equivalent from the point of view of natural laws. The laws are the same everywhere.

This is the belief of a physicist, which he cannot do without. The belief is not quite baseless. Astronomers have observed faraway stars and galaxies through their telescopes. The motion of these faraway objects are well explained by the same laws that govern the motion of the stars and planets that are nearby. If however that were not true, the physicist would have said — 'the fault is mine for not being able to find the correct law of nature'. Even then he would not have had any doubt about the universality of natural laws.

German mathematician Emmy Noether showed that if we consider this concept to be sacred then the conservation of momentum could easily be proved. This may sound fantastic. But the fact that we have such fantastic events every now and then, keep science lively and beautiful.

Similarly, nature has no preference for any given direction. If that were not the case then the momentum of a particle moving due south would have been different from the momentum of the same particle moving due east with the same velocity. Perhaps we would have said — the rest mass of a particular object is 2 gm when moving towards east and 3 gm when moving towards south. But that does not happen in reality. The rest mass remains the same, irrespective of the direction of the velocity. In the same way, we can find the conservation of angular momentum from this observation.

There is one more thing. The time is also homogeneous. This means that the result of an experiment performed today would not differ from the result of an identical experiment performed three million years ago. And the same would happen five billion years later. Nature is not partial to a particular point of time either. Noether showed that this fact easily gives rise to the conservation of energy.

Another name for impartiality is symmetry. If, from the center of a square, we look at the points on its sides we shall see that some points are closer and some are more distant. For example, in the top square of Fig. 1.8, A is more distant from the centre compared to the point

B. In other words, all the points are not equivalent. But if we have a circle instead, all the points on its perimeter are equidistant from the center. These points are therefore identical. So we can say that looking from the centre of a circle all points are the same — there is no preference for any direction. Turning this around, we can say that a circle has 'rotational symmetry'. Meaning, if we close our eyes and someone rotates the circle about its center then we would not be able to discern the difference after opening our eyes. If there is nothing written on a compact disc, then we would not have any means of knowing whether it has stopped at the same point after rotating. The only way to tell whether the same point has come back is to mark the point somehow. But the moment a point is marked, a preference is set. And consequently the rotational symmetry is destroyed. A square does not have this rotational symmetry. That is why we see in Fig. 1.8 that the shape of the square has changed after rotation, whereas nothing has happened to the circle.

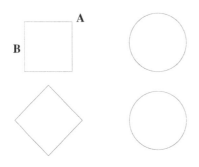

Figure 1.8: A square does not have rotational symmetry, whereas a circle does.

Similarly, we can talk about the homogeneity of space by saying that the space has 'translational symmetry'. The same story. If the magical monster moves the universe to another point in space we would not know about it. This gives us the conservation of momentum. Similarly, time also has translational symmetry — which gives us energy conservation. And the rotational symmetry of space gives us the conservation of angular momentum.

In general, we could say that every conservation law has an underlying symmetry associated with it. Alternatively, a symmetry always gives rise to a conservation law. Exactly that's what Noether said in 1918. Since then the concepts of 'symmetry' and 'conservation law' have become synonymous.

Consider this. We had obtained the momentum conservation from Newton's laws for the first time. Later the limitations of Newton's laws have been discovered and we have learnt that Newton's definition of momentum needs to be modified. We also had to add the momentum of the fields, and to remember that the kinetic mass depends on the velocity of an object. Even after all these corrections, the conservation laws hold good. Perhaps someday we would find the limitations of Einstein's theory, or certain weaknesses of Maxwell's field theory. But whatever new theories may come, if the concept of translational symmetry of space is not discarded then we shall always find a quantity which would be conserved. And this quantity would be known as momentum then.

We have already said that a physicist must have faith in symmetries. If it so happens that the laws he discovers here are not applicable at another point in space, then such laws are completely useless to him. If the laws of nature change from time to time, then what use is all the hard work to uncover them? This is why the conservation laws are so important. This is why the role of symmetry in the drama of physics is so magnificent.

The faithful say that only faith can find God. We shall end this chapter by an example of what faith, in conservation laws, has given science. In the beginning of the twentieth century it became known that the nuclei of certain elements decay spontaneously and electrons come out of them. But sum of the momenta of the electron and the nuclei did not equal the momentum of the original nuclei.

Some people said, get rid of such conservation laws. But Wolfgang Pauli said that the conservation law is correct. It is not just an electron, another particle is also being emitted. We are not being able to observe these particles in our detectors. These were very difficult to detect. They were named neutrinos and it was said that the conservation law would indeed be obeyed if neutrinos were included in the calculation.

Honestly, it was a rather elusive particle. Twenty-seven years after the proposal of Pauli, after a lot of hard work, those particles were detected.

What could be a bigger success for a conservation law?

Chapter 2

Waves and particles

2.1 From Democritus to Dalton

Consider medical science. Checking a person up from outside, feeling his pulse or listening to the nature of his cough — it is quite possible to diagnose certain diseases and even treat them. However could a doctor be satisfied with just that? He cannot and indeed he should not. He needs to know the details of the mechanism working inside a human body — from the location of the brain or the liver to the functioning of the lungs — only then would he be able to treat a serious condition.

Towards the end of the nineteenth century the situation with physics was somewhat like the first kind of medical practice. Meaning, all that we could perceive through our senses — all that we could see, could hear or touch with our hands — were satisfactorily explained by the physicists. But as for the main constituents of matter, the mystery lying hidden beyond our vision — the ideas were not very clear about those.

However, the idea that matter is made up of tiny indivisible particles was in existence even in the ancient days of Kanad or Democritus. To them, it was not very clear how small these particles were, what were their properties or why should they be indivisible. Even now, we do not claim that we have known the last word on the subject. Therefore, it is not very surprising that Democritus' theory was not readily accepted by everyone in 500 BC.

Science went through dark ages for the next 1500 years or so. Then in the fifteenth and sixteenth centuries came the Renaissance, the new awakening. It was a new beginning for the sciences too.

To explain various chemical reactions the chemists had to take recourse to Democritus' atomic theory. The scientists found quite a few interesting facts looking at the results of a large number of chemical experiments conducted in the seventeenth and eighteenth centuries. These were sifted through in the early part of the nineteenth century by scientists like Dalton, Gay-Lussac, Avogadro, and Cannizzaro.

The result of that exercise was the establishment of atomic theory. John Dalton, in the year 1803, declared that — all matter is composed of a large number of tiny, indivisible particles. The name of these particles was *atom*. There are as many different kinds of atoms in the world as the number of chemical elements. All atoms of a given element are the same. Atoms combine to make molecules and when many such molecules come together we have perceivable matter.

The physicists liked this concept. They calculated the pressure, the heat conductivity and various other properties of gases assuming a gas to be made up of a large number of atoms. Soon enough these endeavours also met with success.

It was observed later that the atoms are not really indivisible. They contain smaller particles like the electrons, protons etc. Nevertheless, the concept of elementary particles making up matter became firmly established.

2.2   Newton, Huygens

The other constituent of the universe, besides matter, is energy. We see many different manifestations of energy. Light, heat, sound, electricity — all of these are different forms of energy. The question is, what would be the underlying mystery of energy like we have particles in the matter? How is the energy transported? You light a candle in the middle of a room and light falls on the walls. How did it travel? One end of an iron rod is held against fire and you feel the heat at the other end. How?

The early scientists did not try to answer this question. Perhaps they did not even formulate the question. It was raised for the first time towards the end of the seventeenth century, in the days of Isaac Newton. Newton himself said that light, like matter, is composed of minuscule particles or *corpuscles* of light. When we see the light from a candle burning in the middle of a room falling onto the wall, it means that many such light particles are traveling through the air and hitting the wall. These particles are so small that it is impossible to see them, even more difficult to measure their properties. We perceive them simply because their visibility is an effect of the combination of a large number of such particles.

That is good. But we need to explain all the known properties of light using this concept. Otherwise the theory has no use.

Alright. Firstly, light travels in a straight line. And that is to be expected. These tiny particles, on which no force is acting, should not move in any sort of curved path. They would travel in a straight line, with an unchanging velocity. Newton's laws of motion have taught this to us. If an opaque object is kept in the path of light, the particles would not be able to penetrate that, and a shadow would be created behind the object. Therefore in this context the corpuscular theory of light is successful.

Secondly, light gets reflected from plane surfaces. Exactly like the way a tennis ball rebounds from the surface, the light corpuscles also rebound from a plane surface. The result of this is reflection.

Then the third issue is that of refraction. That is the phenomenon of bending when light travels from one medium to another. This is why a pencil, partially submerged in water, looks bent. If a ray of light comes from a particular direction then the amount of bending would depend on a particular property of the medium, known as refractivity. Newton explained refraction using his corpuscular theory. And he showed that the refractivity is nothing but the ratio between the velocity of light in the given medium to the velocity of light in vacuum. Experiments revealed that refractivity is greater than one in every medium used. The conclusion is that the velocity of light in vacuum is smaller than that in any material medium. Of course, light travels exceedingly fast. There was no way one could have measured that velocity in those days. Even then, a theoretical explanation became available. Therefore, it was again a success story of the corpuscular theory.

However in 1687, within a few years of the appearance of corpuscular theory, a completely different proposition was put forward by the Dane scientist Christian Huygens. He said, light propagates like a wave. Now the word wave evokes the picture of waves in water. In physics, though, the word is used in a larger context. Let us say that some physical parameter is changing over space and time. And the change is such that after a certain time and distance the values are repeated. That is if one looks at the values at a given time then it would seem that after a given distance they are being repeated. We would find similar repetition over time if the values are noted at a particular point in space. When such a thing happens we say that the parameter has a wave behaviour. In water waves the height of the surface is different at different points, then again it is different at different times at the same spatial point. Likewise, if we look at the variation of the temperature then we could say that a wave of heat is propagating. Or in the case of pressure it would be a wave of elasticity and so on.

However, it was not quite clear to Huygens what was the parameter which varied during light propagation. Neither was it easy to explain the phenomenon of straight line propagation, reflection or refraction. However, Huygens persevered and proved all of these from a wave-theory viewpoint. The question of straight line propagation was hardest of all. Still he managed to prove it.

Yet, the scale was tipped towards the corpuscular theory. Remember the famous saying of Archimedes — if there exist two theories that can explain the same phenomenon, the simpler one should be correct.

And there was one other issue. The influence of Newton in the community of mathematicians and physicists was so strong that scientists wanted to show their loyalty towards Newton against the lesser known Huygens. Moreover, there was no extra usefulness in Huygens' theory.

Wait, wasn't there anything more advantageous at all? In 1669, Erasmus Bertholinus had observed that for certain materials an incident light wave gets refracted in two different directions. A ray of light gets divided into two. For the first time Huygens explained this phenomenon of double refraction using his wave theory. The corpuscular theory had no explanation for this. Even though people believed that the explanation surely existed. It just needed someone to come up with a proof. And they continued to accept the corpuscular theory.

However, a bit of confusion still remained. The concept of refractivity obtained from Huygens' wave theory turned out to be inverse of that obtained using Newton's wave theory. It was found that refractivity is the ratio between the velocity of light in vacuum to that in a material medium. Then according to the wave theory the velocity of light is maximum in the vacuum. There was no way measure the velocity. But it is clear that one of the theories would have to be incorrect. Which one?

2.3   Young, Fresnel

In his later years Newton himself conducted an interesting experiment. He placed a glass lens on top of a plane mirror with a point source of light above, from which a monochromatic (of a single frequency) light was being emitted.

The source was placed such that after being bent through the lens the parallel light rays fell upon the mirror surface perpendicularly. Therefore, the light rays were reflected from the mirror and returned along the same path as has been shown in Fig. 2.1.

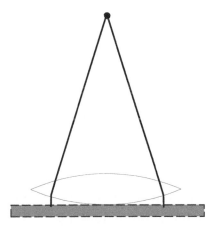

Figure 2.1: Experimental arrangement for producing Newton's rings.

Now the pattern seen beneath the lens has been shown in Fig. 2.2 — dark circular disc in the centre, surrounded by a bright ring, surrounded by a dark ring and so on.

Where are these dark sections coming from? Very difficult to explain this through corpuscular theory. Huygens was dead by this time. So, nobody really bothered to look at its wave aspect at all.

The explanation had to wait for almost a century. Other similar phenomena were observed by scientists like Thomas Young and Augustin-Jean Fresnel in the beginning of the nineteenth century. And they realised that these could be explained nicely — not by corpuscular theory — but by wave theory.

Figure 2.2: Newton's rings — interference rings observed by Newton.

What is the explanation then? For the ease of under-standing let us use the water waves again. Say we drop two stones in water. Both the stones would give rise to waves. Now what would be the situation at a point where both the waves are arriving? The answer is that the height of wa-ter surface would be the sum of the heights that it would have attained due to the two waves separately. A situation like this, created due to the superposition of the waves, is known as *interference*.

If a particular point on the water surface is supposed to rise by 3 cm because of one wave and is expected to drop down by exactly 3 cm due to the other then the net change in the height of the water surface as a combined effect of the two waves would be zero. The amount of light received at a point in a similar situation would also be zero. That is, it would be dark there. Then again at some other point the resultant light would be brighter than in the case of one single wave. And it does happen.

In Newton's experiment the light coming directly from the source interferes with the light reflected from the mirror below. As a result there is bright light at certain points and darkness elsewhere. Detailed calculations precisely showed the positions to expect bright light and darkness. The calculations matched the experimental results perfectly. Many more such experiments on interference were conducted by Young, Fresnel, Arago, Malus etc. Everywhere the experiments were successfully explained by the wave theory. Many other phenomena were also discovered besides interference and were explained satisfactorily by the wave theory.

Therefore it really was time to say goodbye to the corpuscular theory. Because all the optical phenomena were being explained by wave theory and not by the corpuscular theory. Only one question still remained as far as refraction was concerned.

That puzzle was also solved by scientists like Fresnel and Fizeau. They conducted experiments to measure the velocity of light and showed that the velocity of light is always smaller in a material medium than in vacuum.

This then established the supremacy of wave theory of light unquestionably.

2.4 Maxwell

James Clarke Maxwell did not really work on the nature of light. His research focused on electromagnetism — i.e., phenomena related to either electricity or magnetism or both. Before this, Michael Faraday had shown that it is not possible to describe either the electric or the magnetic phenomena on their own. A magnetic field varying with time gives rise to an electric field — the principle which is used in a dynamo. Maxwell himself realised that the reverse is also true, that is a changing electric field generates a magnetic field as well.

Maxwell constructed a theory combining the enormous volume of experimental results compiled by Faraday and his own understanding of these. This theory could explain all the electromagnetic phenomena.

It became clear from Maxwell's theory that real physical quantity is the electromagnetic field. An electric charge or a current flowing through a wire or a permanent magnet generates an electromagnetic field around it. The electromagnetic energy propagates as a wave in this field.

Meaning we have the wave theory here again. But what is the proof of that? Well if there is wave there should be interference. In 1887 Heinrich Hertz showed the existence of interference in the context of electromagnetism.

We already know the definition of waves. Therefore we need to know what is the physical parameter whose variation gives rise to the electromagnetic wave. It is known as the intensity of the electric field. Of course, the intensity of the magnetic field shows similar variation and we can also think of the electromagnetic wave from this point of view.

Now it was observed that the velocity of this electromagnetic wave is exactly equal to the velocity of light. Maxwell concluded from this that light is simply a special form of the electromagnetic wave. Therefore, the question that remained unresolved since Huygens' time — the parameters whose variation gave rise to the light waves — was also answered now. It is nothing but the intensity of the electric field, or alternatively of the magnetic field.

However, after all this Maxwell made a curious mistake. Actually a wave reminds us so much of water waves that we do not feel comfortable without identifying something whose surface goes up and down as a result of wave propagation. If that is the case then what is this "something"? Is it some kind of material medium? But how is that possible? We know that light travels through the vacuum in space, from distant stars to our Earth. So Maxwell said that there exist a medium called "ether" which is all-pervading. Even interstellar space is not empty but is full of ether. Light waves propagate through this ether.

The theory of ether died at the hands of Einstein. And therein lies the beginning of the theory of relativity. But that is another story altogether. Let us continue with the line of reasoning followed till now. So, there may not be material in space but electromagnetic fields do exist. And the intensity of that field varies with space and time — constituting a light wave. Light propagates in the form of this wave.

Not only light, sound also propagates like a wave. In other words, every form of energy propagates like a wave.

So physics reached a very satisfactory high point. On the one hand there is matter — all of which can be explained through atomic theory. And then there is energy which has its basics laid down in the wave theory. Quite a beautiful scenario. Lord Kelvin, one of the most eminent physicists of the nineteenth century, even went on to say that all the fundamental rules of physics had been discovered, the only task remaining was to successfully apply them to more and more complex contexts. In addition, one required technologically advanced devices to measure fundamental parameters more accurately.

The story was somewhat like this at the end of nine-teenth century.

2.5   Hertz, Lenard

Every once in a while history makes serious fun of hu-mankind. The experiment of Hertz which definitively proved the correctness of Maxwell's theory, helped to build the mansion of classical physics, itself contained the seeds of destruction of that mansion. The situation is like this. Hertz constructed a small instrument to observe the inter-ference of electromagnetic waves like we observe the inter-ference of light (perhaps sometimes we use an instrument to magnify the effect but in the end we do see it with our eyes like in Fig. 2.2). A wire is bent and the ends are brought together without the ends actually touching (like it is seen if Fig. 2.3). If an intense electromagnetic wave now comes near the gap a spark would be seen across it. If there is no electromagnetic wave there would be no spark. Using this construction Hertz could show that as a result of the interference between two electromagnetic waves at certain points there would be strong sparks, at some places weaker and no sparks at all at yet other places. This means that because of interference the intensity of the electromagnetic field is increasing at some points, decreasing at others and completely vanishing at certain points. This is what New-ton observed in case of light.

Alright. But Hertz observed that the sparks are brighter if light is incident on his instrument. A faint glow became discernible even in places where there were no sparks ear-lier.

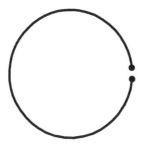

Figure 2.3: Hertz's instrument for observing the interference of electromagnetic waves.

Why? An explanation was developed by analysing the situation continuously. Some electrons are freed from the wire when light falls on it. Due to this an electric current is being produced since the electrons carry charge. This current, being generated due to the incidence of light upon metal, is now known as the photoelectric current.

The person who investigated this photoelectric current in detail was Philip Lenard. And the results of his investigation turned out to be the cause of many a scientist's headache.

Because Lenard noticed some peculiar phenomena. Every light wave has a frequency. The number of variations in the field intensity per unit time is called the frequency. Lenard found that if the frequency of the incident light is smaller than a particular value no photoelectric current is generated.

Lenard also noticed that the velocity of the ejected electrons does not depend on the amount, i.e., the intensity of the incident light.

Both are quite perplexing from a wave theory viewpoint. Because the main hypothesis of the wave theory is that the energy of a wave is not concentrated at any one point. Such concentration is necessarily a property of particles. When a wave is formed in water due to the dropping of a stone, its energy is distributed at every point of the wave.

Similarly, when a light wave is incident upon a metal, the electrons inside would absorb the energy from various parts of the wave. Finally they would have enough energy to come out of the metal. If that is the case then the brighter the light, the more energy there is at any point of the wave. Consequently, the total energy absorbed by an electron would be larger for a brighter light and it would have a higher velocity. And frequency should have no role to play according to this picture. Then again the electrons would require some time to collect and an adequate amount of energy to be ejected. But Lenard showed that the current started flowing as soon as the light was incident upon the metal.

Thus entered the dark shadow in the happy house of physics. The publication of the results of Lenard's experiments in 1902 created a stir in the scientific community. The resolution of this problem came three years later. But we shall have to go back in time, in order to search for the roots of this solution.

2.6 Planck

The topic of our discussion in this section is not light but heat. Another form of energy.

There are three ways in which heat can propagate. Two of them require the help of material media and are called *conduction* and *convection*. In the first method the atoms of the medium do not suffer any displacement — one atom simply transfers heat energy to the next one. Something like a long line of men, where the first person takes a brick and transfers it to the second one, he then in turn gives it to the third and so on. In this way the brick itself travels but the men retain their positions. This is conduction. But if it happened in a different way, say a person transports the brick himself from one point to another — then we would have a case of convection.

But there is another way to transport heat which does not require any help from a material medium. This method is known as *radiation*. The heat comes to the Earth from the Sun or a star via radiation because it has to cross outer space which is devoid of any material medium. There is no other way of heat transport without medium.

Confusion started with this radiated heat. Say a black body has been heated to a particular temperature. In turn this object would start radiating heat. And the heat would come out in the form of waves. For the moment, let us not worry about the definition of this black body according to physics. We would rather like to know how much energy is radiated by this black body at a particular frequency.

The branch of physics dealing with energy till then is called *thermodynamics*. Lord Rayleigh hinted at the way this branch could be used, in conjunction with the laws of classical physics, to arrive at an answer to this question. The solution was obtained by James Jeans. But the theoretical results differed significantly from experimental observations. In 1896 Wilhelm Wien tried a different tactic, extending the tenets of classical physics with some new assumptions. His results were not as disastrous as Rayleigh and Jeans. But they didn't quite match the experiments either.

When such a stalemate is reached there exists no other way except to shake the foundations themselves. Max Planck did exactly that. Till now it was believed that an object can emit or absorb any arbitrary amount of energy. Similarly any arbitrary amount of energy can be transported from one place to another. But in a meeting held in 1900 Planck announced that when energy is emitted by a black body it comes out in small packets. Small packets or *quanta* of energy. Each of the packets with frequency f would have the same amount of energy given by hf. This h is a fundamental constant. In honour of Planck we now call this the Planck constant.

So we see that heat at a frequency f cannot be emitted by any arbitrary amount. It could be emitted in one or two or three packets but certainly not in one and a half or three and a quarter packets. Therefore, the radiated energy is always equal to hf or some multiple of that — any deviation from this is not possible.

The results of Planck's calculations, made with this assumption, matched the experimental results exactly.

That is alright, but what exactly do we have here? Doesn't it sound quite like the corpuscular theory again? Whether you call it quanta of energy or something else — there's no way to hide the truth — those quanta do have particulate behaviour. But if we believe in the wave theory we have to assume that any arbitrary amount of energy can be absorbed or emitted by matter and we would be back with the results obtained by Rayleigh or Wien.

Planck himself was in favour of combining both and therefore chose a middle path. He said that energy is actually continuous. Only when it is emitted from or absorbed by a material object it happens in quanta. Thereafter it again reverts to its continuum nature.

Isn't this quite like having the cake and eating it too?

2.7   Einstein

It goes without saying that nobody accepted Planck's proposal happily. Planck himself was the unhappiest. He lived for another forty-seven years but for the rest of his life he resented the very quantum theory he helped build. Yet, it has been agreed by scientists since 1924-'25 that it is Planck's hypothesis which gave birth to modern physics. In any case, let us get back to our discussion of history - the history which is at a crossroads now.

Right at this moment of transition there appeared Albert Einstein in the glow of the footlight. When everyone was raising their eyebrows over the proposal that energy quanta revert back to continuum immediately after emerging out of the medium, Einstein said there was no need to take offense with this concept. In fact, the discomfort arises because it is contrary to our common sense. But then what is common sense? "Common sense is the collection of prejudices acquired by age 18."

So there is no point in saying such things. Whether Planck's proposal is correct or not cannot be judged philosophically. Was it possible to say whether the wave or the corpuscular theory was correct in Newton's time? It had to be applied in different situations and only then it could be resolved.

What was the solution then? Planck applied his hypothesis only for heat radiation, it now needed to be applied to many different situations. Then if required the hypothesis would have to be modified and restructured.

By this time, in 1902 the results of Lenard's investigations on photoelectricity have been published. We have already said that these could not be explained by the wave theory. Let us see now whether the light quanta could save the situation.

Let us say we have an open water tank with some fish playing in it. Now a stone is thrown into it. The stone has some kinetic energy which would get transmitted to the water waves. As the waves hit the fish they too would receive some of that energy.

The amount of energy each fish received would be a small fraction of the energy the stone originally had. So if we see that a single fish received so much energy that it was thrown out of the water like a bullet, then we would really be very surprised. Yet exactly that is what is happening in the case of photoelectric current. The electrons are like the fish and the metal is like the holding tank. Light falls on the metal and the electrons jump out.

What if we want to forget about the wave completely? If we say that the stone is directly hitting the fish bringing in its kinetic energy then it would not be surprising to see a fish being thrown out. It does not usually happen in the case of a fish in water, but that really is the situation for photoelectric current as proclaimed by Einstein.

Well, the comparison is nice but there is a small confusion. What is the analog of the stone here? Light? Does that mean the light is coming in quanta even in this case? Planck said that light is only emitted in quanta. But the metal is absorbing instead of emitting light here.

Einstein said, yes. When a material radiates energy it emits it in separate quanta. Similarly the absorption too happens in units of energy quanta. And light of frequency f only gets absorbed in quanta of energy hf each, or as particles with energy hf.

Therefore, when a light of frequency f falls on a metal it is absorbed as quanta of energy hf. These quanta (or particles) enter the metal and collide with the electrons. An electron can receive some energy through this collision. How much? At most hf, because a quantum of light does not have more energy in it. The electron becomes energetic as a result and would try to emerge out of the metal. However, there is something like a wall it would have to scale in order to cross the boundary of the metal. Let us assume the energy required to scale this wall is w. Then the energy left with the electron would be $hf - w$ and it would come out with this amount of energy. Then what would be the maximum kinetic energy of the ejected electron? If we call this maximum energy T then,

$$T = hf - w.$$

And if the electron receives an energy less than hf then its energy would be less than this.

Simple yet elegant, quite extraordinary! Einstein received the Nobel prize for his 1905 explanation of photoelectric effect, not for the theory of relativity.

All of Lenard's research could be explained by this equation of Einstein's. To begin with, if the light has a frequency such that hf is smaller than w then the electron can never scale the boundary wall. Therefore, it would not do to have a very small value of f. At the least it has to be greater than w/h. Secondly, there is no mention of the brightness or intensity of light in this equation. Therefore, the kinetic energy and accordingly the velocity of electrons would not depend on the intensity of light. Thirdly, it is quite obvious why the electrons are ejected as soon as light falls on the metal. Does it take long for the ball to fly off after it has been hit by a bat? There is no question of accumulating energy little by little, this is a case of one time hit.

Do we then have to go back to the corpuscular theory of Newton? But there is no way of denying the fact that only wave theory can explain the phenomenon of interference. And not only interference, there exist many such similar phenomena that we have not even mentioned.

Einstein modified Planck's statement a little. When the energy flows on its own it behaves like a wave. But when it comes in contact with matter its particulate nature prevails.

2.8   Bose, Bohr

The theory remained like this for a while despite a nagging bit of unease. Though Planck's quantum hypothesis was now being applied to various situations. Einstein himself used it to calculate the specific heat of solids.

In 1924, Satyendra Nath Bose did something wonderful. Let us say there is some radiation confined within a box. Remember the story in which a king put his chest out in the sunlight during the day and then brought it back full of heat. Something like that. Of course, the radiation would escape through the sides of the king's chest. But our box has mirrors on its sides. When radiation falls on the sides it gets reflected back. Now this box has a tiny hole in it — so tiny we cannot even conceive of it. The radiation coming out of this hole is exactly like the radiation from a black body. Satyendra Nath considered this radiation to be made up of a large number of quanta. And his calculations resulted in the same formula as Planck's which could not be found by Wien or Rayleigh.

Then how can we say that energy shows its particle nature only when interacting with matter? It looks like the energy behaves always like particles. However it is not possible to throw away the wave theory altogether. So what is energy — wave or particle?

It is both wave and particle — said Niels Bohr and his associates, which is quite contrary to our common sense. They said there is nothing that prevents something with wave nature to show its particle nature in certain situations. Energy has a dual nature. Sometimes its behaviour can be explained by the wave theory and other times by the corpuscular theory. To understand the nature of energy we need to know both of these.

2.9   de Broglie

The situation then became slightly different than in the time of Maxwell. The properties of matter indicate that it only has particle nature. Whereas the energy has both wave and particle nature.

Louis de Broglie, however, protested. He was a veritable aristocrat. Physics was a kind of hobby for him. In 1924 he declared, in quite a lordly fashion, that it is not possible. Agreed, energy has a dual nature. But then the matter would have dual nature too. There must exist certain properties of matter which require wave theory to be explained.

And the frequency of that wave could also be calculated using Planck's formula. That is, if the energy of the material particle is E then the frequency f of the associated wave would be given by $f = E/h$.

Everyone thought this was quite insane. Certainly all the properties of matter that we know of can be explained by its particle nature!

Really? The electrons inside the atoms can never have any arbitrary amount of energy. Meaning their energy is not continuous. Electrons inside an atom can only have certain definite amounts of energy. Within a few years of de Broglie's proposal Erwin Schrödinger derived an equation for the wave of a particle. And he showed that it is possible to explain this situation through his equation.

This had been a headache for the scientists till then. Though Niels Bohr had suggested a solution but that was almost like trying to work out the method of a problem after knowing the answer to it. Of course, that was not an easy task either. However there were many other important aspects of Bohr's theory. This was the first emphatic statement about the need to abandon the concepts of classical physics. But we shall discuss that later. There is no denying the importance of Bohr's theory, but the entire scientific community had been waiting for something like the work done by Schrödinger.

For the hardcore argumentarians, who were still not happy with this, there came the results of two experiments in 1927. Both the experiments were basically the same. One was performed by Clinton Davisson and Lester Germer and the other by George Thomson. They managed to *diffract* the flow of electrons.

We have not used the word diffraction till now. Diffraction is nothing but a type of interference. We know that interference is a property of wave. So it is obvious that the flow of electrons is behaving like a wave.

So we see that there are material properties which can only be explained by the wave theory. Not only energy, but matter also has a dual nature then.

2.10 Jekyll, Hyde

Alright, whatever we may or may not understand, one thing is amply clear — the history of physics is full of dramatic events. The clash between the particle theory of Newton and the wave-hypothesis of Huygens; the undoubted victory of the wave-hypothesis in the Young-Fresnel era; the demonstration of the relation between electromagnetic waves and light by Maxwell; then Hertz's experiment like 'The Dramatic Irony'; the failure of wave hypothesis to explain the radiation of heat and its subsequent explanation by Planck's hypothesis; the new insight of Bose for heat radiation; Bohr's declaration regarding the particle-wave dual nature of energy; de Broglie's claim for the dual nature of matter as a result and finally the direct proof of that dual nature — almost each and every one of these steps made for dramatic confrontation!

However, like a modern drama we have not yet said the last word. Suppose today someone asks a physicist — 'See here, a lot has been said, but could you give me the answer to a simple question? What is light exactly, particle or wave? Or take matter, for example, an electron — what is it? A particle or a wave? Which one?'

A physicist would smile and say — 'Light is simply light. It has certain properties. On the other hand, an electron is just that, an electron. You have already decided that a particle behaves this way and a wave thus. But neither light nor electron can be classified according to such concepts. For example, you have decided to call a person ethical if he feels socially responsible for the downtrodden, does not take bribes, etc. Now you find someone who has serious concerns about the poor but takes huge bribes from the rich. Then you have a problem to decide whether or not to call this person ethical.

The situation is similar for the light or the electrons. For certain events they would behave like waves, in some others like particles. This wave-particle duality of matter and energy is an inherent concept of quantum theory, or equivalently of, modern physics. These are neither particles nor waves but both particles and waves at the same time — like Dr. Jekyll and Mr. Hyde.'

Chapter 3

Insider's story of the atomland

3.1 Dalton

We have already heard about atoms. Continuing along the line of ancient Greek scientists and considering the wealth of experimental results obtained in the post-Renaissance phase, Dalton said that everything must be composed of atoms. An *element* consists only of one kind of atoms. For example, take iron. All iron atoms are exactly alike. Similarly, metals like aluminium, copper, gold, silver, zinc or gases like hydrogen, nitrogen, oxygen or materials like diamond, mercury, sulphur — all of these are elements. Ninety-two such elements have been found in nature.

Everything else are *compounds*. A certain number of atoms of a particular element combine to form a molecule of the same element. The rust formed on a piece of iron is a compound — the molecules are formed when iron atoms combine with the atoms of oxygen present in the air. A single molecule of edible salt has one atom of sodium and one atom of chlorine. The chemical symbol of sodium is Na and Cl stands for chlorine. So a salt molecule is represented by NaCl. Even water is a compound, symbolically written as H_2O — each molecule containing two hydrogen atoms and one oxygen atom. In this fashion all the material of the world is composed of the aforementioned ninety-two kinds of atoms.

Not only did we understand the basic building blocks of matter with the advent of atom. The mystery of chemical reactions became clear too. Dalton said that in a chemical reaction neither are the atoms destroyed nor new ones created — they simply change partners. Caustic soda contains sodium, and hydrochloric acid has chlorine in it. When these two are mixed the sodium and chlorine combine to form salt, NaCl, as we have said before. And the rest is simple water. Writing in terms of chemical symbols this reaction looks like the following,

$$\underset{\left(\substack{\text{Caustic}\\\text{Soda}}\right)}{\text{NaOH}} + \underset{\left(\substack{\text{Hydrochloric}\\\text{Acid}}\right)}{\text{HCl}} \rightarrow \underset{\text{(Salt)}}{\text{NaCl}} + \underset{\text{(Water)}}{H_2O}$$

Of course, all reactions do not proceed with such ease, external heat or pressure need to be applied in many cases. But the basic message is simple — a chemical reaction takes place when the partner of one goes to another.

Therefore chemistry became much easier now. It was no longer necessary to sweat analysing millions of compounds. If one understood the nature of these ninety-two elements, the way they combine with each other, then the key to understanding the chemistry has already been found.

1	2	3	4	5	6	7	8	9	10	11	12	13	14	15	16	17	18
1 H 1.008																	2 He 4.002
3 Li 6.94	4 Be 9.01											5 B 10.81	6 C 12.01	7 N 14.01	8 O 16.99	9 F 19.00	10 Ne 20.18
11 Na 22.99	12 Mg 24.31											13 Al 26.98	14 Si 28.09	15 P 30.97	16 S 32.06	17 Cl 35.45	18 Ar 39.95
19 K 39.09	20 Ca 40.07	21 Sc 44.96	22 Ti 47.90	23 V 50.94	24 Cr 51.99	25 Mn 54.94	26 Fe 55.85	27 Co 58.93	28 Ni 58.70	29 Cu 63.55	30 Zn 65.38	31 Ga 69.72	32 Ge 72.59	33 As 74.92	34 Se 78.96	35 Br 79.90	36 Kr 83.80
37 Rb 85.47	38	39	40	41	42	43	44	45 Rh 102.9	46 Pd 106.4	47 Ag 107.9	48	49	50	51	52	53 I 126.9	54 Xe 131.3

Explanation of chemical symbols

Ag : Silver	Al : Aluminium	Ar : Argon	As : Arsenic	B : Boron	Be : Beryllium
Br : Bromine	C : Carbon	Ca : Calcium	Cl : Chlorine	Co : Cobalt	Cr : Chromium
Cu : Copper	F : Fluorine	Fe : Iron	Ga : Gallium	Ge : Germanium	H : Hydrogen
He : Helium	I : Iodine	K : Potassium	Kr : Krypton	Li : Lithium	Mg : Magnesium
Mn : Manganese	N : Nitrogen	Na : Sodium	Ne : Neon	Ni : Nickel	O : Oxygen
P : Phosphorus	Pd : Palladium	Rb : Rubidium	Rh : Rhodium	S : Sulphur	Sc : Scandium
Si : Silicon	Se : Selenium	Ti : Titanium	V : Vanadium	Xe : Xenon	Zn : Zinc

Table 3.1: First four rows of the periodic table, and a few elements in the fifth row.

3.2   Mendeleev

So far we have talked about the scenario as it was in the early nineteenth century. Mendeleev explained the situation to be even simpler in the latter half of the nineteenth century. Some of the elements behave in an uncannily similar fashion. This was already being suspected but to make it clear Mendeleev used a table. We know this as the *periodic table*, the first few lines of which have been shown in Table. 3.1.

The meaning of this table is very simple. All the elements on the same vertical column have similar properties. Consider, for example, sodium and potassium. Caustic soda is a strong base. Whereas, caustic potash, a similar compound made from potassium is again another strong base. Both of these are typically used to clean surfaces with stubborn stains like the kitchen work-bench.

The similarity between fluorine, chlorine, bromine, and iodine is equally spectacular. Their behaviour in chemical reactions are the same, they form similar compounds. Xenon, argon, neon, and helium again have similar properties. These gases are so inert that they do not want to react with any other element. This is the reason these are used for gas-discharge lamps (in which light is produced by passing electricity through a gas) to prevent accidents. Because most other gases are prone to chemical reactions at such high temperatures. Similarities like this can be mentioned about all the other columns in Table. 3.1.

The table may look simple, but in 1869 it was not an easy task to put it together. Firstly, how do we know the number according to which the elements have to be ordered in the table? Mendeleev had ordered them roughly according to their atomic weight. The atomic weight of each element has been mentioned at the bottom of each box in Table. 3.1. For example, the atomic weight of magnesium is 24.31, which means that one atom of magnesium is 24.31 times heavier compared to one atom of hydrogen. Same with other elements. Therefore, from Mendeleev's table we find that even though the atomic weight keeps increasing the chemical properties of elements have some sort of a repetitive nature. In short, this fact was the path-breaking discovery of Mendeleev.

But getting here was not and easy task, as we have said before. Consider beryllium for example. Its atomic weight was thought to be about 14. Then in the second line of Table. 3.1 the elements should appear in the sequence — lithium, boron, carbon, beryllium, and so on. In that case, boron and magnesium appear in the same vertical line, aluminium comes directly below carbon and beryllium is positioned right above silicon. But silicon has no similarity with beryllium. And boron has nothing in common with magnesium but is similar to aluminium. What happens now? Mendeleev said — the atomic weight of beryllium must be closer to nine. Better measurements did indeed find it to be so. This can be seen in the table. Mendeleev was absolutely right.

The next difficulty came after zinc. The existence of gallium or germanium was not even known at the time. Yet, Mendeleev had serious objections against placing arsenic right after zinc. Because that would place arsenic directly below aluminium, whereas the properties of arsenic resemble those of phosphorus. With unbelievable confidence Mendeleev declared that the place of arsenic would have to be directly below phosphorus. As expected, people questioned this decision, asking about the two empty slots. Mendeleev replied that there must exist two other elements which would occupy those places. He predicted a number of properties for the first one, by comparing it with aluminium. Similar predictions were also made for the second element. He even named these two. The first one was called gallium and the second germanium. Not surprisingly both of these were discovered within another twenty years. Their properties were also according to the predictions of Mendeleev. The real importance of germanium, first conceived by Mendeleev, was realised after another seventy years or so, around the middle of the twentieth century. Germanium or silicon (from the same family) is needed everywhere today — from the humble transistor radios to the highly advanced spaceships.

What about cobalt and nickel? Nickel should precede cobalt, given their atomic weights. Yet, cobalt is similar to rhodium and nickel to palladium. Is this again a case of wrong atomic weight, like it happened for beryllium? Not really. Then what could be the reason for cobalt to be in the twenty-seventh place and nickel in the twenty-eighth? Mendeleev could not quite understand the situation. But he refused to budge and forcefully maintained that nickel would have to be the twenty-seventh element and cobalt, the twenty-eighth.

It appeared that there were several other places in the periodic table with similar confusion. History needed to wait for another half a century for the mystery to be solved. And we have to wait for a few more sections to get to that point in our current journey. Meanwhile let us take a look at the status of research on electric current when all this has been happening in the field of chemistry.

3.3 Thomson

What really is an electric current? It was seen that if a bunch of particles, carrying electric charges, move together in a particular direction an electric current is produced. Towards the end of the nineteenth century, Joseph Thomson showed that a particular type of negatively charged particles are responsible for carrying current inside a gas. It is always these same particles which take part in current conduction, irrespective of the type of the gas. Thomson named these *electrons* since they carry *electric* current.

Dalton's atom had no charge. If two such atoms are brought together the net charge would again be zero. Same would happen for ten atoms, a hundred, a thousand, a million or even a billion such atoms. In fact, nothing should be charged if everything is ultimately made up of atoms. Neither can there exist any electric current.

Then where did these electrons come from? Why do these same electrons appear in all the gases? And why ever is the mass of an electron much smaller than that of an atom?

One possible resolution could be that the atoms are not really indivisible even though their name suggests so. They could be made up of smaller components. It was felt that the electrons could be one such component. Thomson encountered electrons in all the gases because every atom contain some electrons.

The electrons are negatively charged. Then for the atoms to be neutral it would have to contain some positively charged particles as well. Those positively charged particles were named *protons*.

Why is it that only electron flow was seen by Thomson in the case of electric current through a gas? Since the protons also carry charge, electric currents would be generated by proton movement too. Yes, that is also possible. However, laboratory measurements showed that the protons are almost 1840 times heavier than the electrons. Therefore much larger energies are required to move the protons. Because of this, it is primarily the electrons which carry current, not only in gases but also in all the electrical wires used for typical domestic purposes.

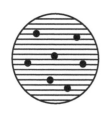

Figure 3.1: Plum-pudding model of an atom.

Cobbling all these concepts together Thomson created a picture of an atom which looks somewhat like that shown in Fig. 3.1. The positively charged particles, meaning protons, form a cake-like structure. And the electrons are stuck on this like the plums in a cake. The mass is largely due to that of the positively charged particles. The characteristic properties of an atom, of course, depend on its mass. Other than that, all the atoms have the same structure. Just that some atoms are like big cakes and some small. The bigger cakes have larger number of plums and smaller ones fewer. The touch of humour is obvious in the name of Thomson's *plum-pudding model* of atom.

3.4   Rutherford

Ernest Rutherford and his collaborators performed an important experiment in the early part of the twentieth cen-

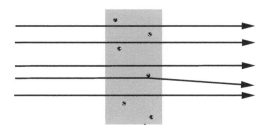

Figure 3.2: Rutherford scattering according to the plum-pudding model.

tury, around 1911. They fired a beam of alpha particles at a very thin gold leaf.

For the moment it is not necessary to know about the alpha particles. It would suffice to know that they are positively charged. While moving through the gold leaf these alpha particles would interact with the atoms and would get deflected in various directions. Rutherford measured these deflections with the help of a detector. And he noted that rather curious results were obtained in this experiment.

Let us try to understand the situation. Say, we have a cake with plums in it. Now we are firing small bullets at this. What would happen? If the bullets have enough energy they would penetrate the cake and come out on the other side. In which direction? Mostly they would move along straight lines. In a few cases the bullets may get a little deflected through close encounters with the plums. However, that would be very rare as seen in Fig. 3.2. If the atoms are compared to the plum-cakes and the alpha particles to the bullets then Rutherford should also have seen something similar.

But what they observed was totally different. They saw bullets getting scattered in every direction. True, most of

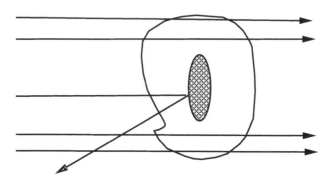

Figure 3.3: Rutherford scattering, as it would have been, for an atom modeled after a mango.

the bullets were going straight, but there were quite a few going in other directions. They even noticed that once in a while some of the alpha particles returned back like a boomerang.

As if the bullets were not thrown at a plum-pudding but instead at a ripe mango. Bullets hitting the soft fleshy part of the mango were going more or less straight. But the ones hitting the hard seed got deflected in all directions, some, of course, straight backwards like in Fig. 3.3.

Rutherford said that the atoms are actually similar to this picture. There exists a compact part in the middle, called the *nucleus*. This part contains positively charged particles, meaning protons. The gold atom is quite heavy, as it has many protons. These are compacted into the tiny nucleus. The alpha particles are also positively charged. These two positively charged objects repel each other. The charge of the gold atom is equivalent to that of 79 protons. Hence the force of repulsion would be very large. If an alpha particle comes very close to the gold nucleus at high speed, the repulsion would be quite strong. This makes the alpha

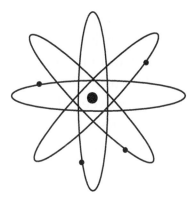

Figure 3.4: The Bohr model of an atom.

particles scatter in all possible directions, including going backwards.

Now, if the size of the nucleus happens to be large, the protons would not be very compactly packed in it. An alpha particle hitting the nucleus on one side would be far away from the protons on the other side. As a result, the force of repulsion coming from those protons would not be strong. The limited repulsion exerted by the nearby protons would still be able to deflect the alpha particles but that would not be very large. The alpha particles would mostly go straight. And we would be unable to explain the results of Rutherford's experiment. A small and compact nucleus is required for this. Calculations showed that the diameter of the nucleus should be about one hundred thousandth of that of the atom. That is, if an atom is like a football field, the nucleus would be a grain of mustard in that field. Yet, most of the mass of the atoms would be concentrated in that grain. The comparison with the mango seed is not very appropriate — the seed is neither that compact nor that heavy.

Rutherford thought about a different analogy: that of the solar system. The nucleus is at the centre like the Sun, with the electrons orbiting around it like the planets, something like the picture shown in Fig. 3.4. There is nothing in the intervening space between the Sun and the planets, nothing but pure emptiness. An excellent comparison, which resulted in the model being known as the *solar system model* of the atom.

3.5   Mosley

Meanwhile, the chemists have also made a lot of progress. They realised that they had to give up another of the basic premises from Dalton era, much like the indivisibility of the atom. It appears that all the atoms of a particular element are not exactly alike. Take lead, for example. Clearly, we have three kinds of lead — one with an atomic weight of 206, another with 207 and yet another with 208. The chemical properties of all three are the same, indicating that they are lead indeed. These were called the *isotopes* (meaning "same place", to indicate that they occupy the same place in the periodic table) of lead. Iron too has three major isotopes, with atomic weights of 54, 56 and 57. Among these 56 is the most commonly obtained natural isotope, the other two are relatively rare. The atomic weights given in Table. 3.1 are average estimates only.

But why do we have these isotopes? And why do we sometimes need to re-position the elements in Mendeleev's table?

Both the questions were answered at the same time, from a completely different viewpoint. They were provided by Henri Mosley, in 1913.

Mosley had been working with the X-rays. When high-speed electrons impinge on a material surface, a type of electromagnetic wave is emitted. These are the X-rays.

What would be the wavelength of these rays? There is nothing fixed about it. There could be waves of different wavelengths. However, rays with certain specific wavelengths were seen to be exceptionally bright. Mosley measured the wavelength of these bright rays and obtained a law concerning them. The law is very simple and is given by,

$$\frac{1}{\sqrt{L}} = a\,(Z - b).$$

Here L is the wavelength of the bright ray, for which this law has been invoked. Now, the brightness is not a characteristic of one particular wavelength. It is seen for a number of wavelengths. The specific wavelength would be determined by a and b. Mosley observed a special bright ray arising out of all the elements for which both a and b are fixed. So specifying Z would then give us the wavelength of the bright X-ray from a particular element.

But what is Z? Ah, we did not talk about this. We said that the X-rays are produced when high-speed electrons are incident upon a material. Mosley's law can only be used if that material happens to be an element. Z denotes the position of the element in the periodic table. This means that Z for cobalt is 27, for nickel it is 28 and 13 for aluminium. Therefore if we take the elements one after another, generate X-rays by making electrons incident upon them, measure the wavelength of the bright X-rays and calculate $1/\sqrt{L}$

— then we'd see that $1/\sqrt{L}$ increases monotonically with increasing Z.

The houses on a city street are numbered for easy searcheability. But the numbers assigned to the elements in the periodic table are not like house numbers. These not only serve the purpose of arrangement but have a deeper significance to them. It is clearly seen that certain properties of an element depend specifically on this number.

What is the meaning of this? Mosley had an answer to that as well. He said the explanation is hidden inside the nucleus of the atom. Cobalt occupies the twenty-seventh place in the periodic table. This means that the charge of a cobalt nucleus is equal to twenty-seven times the charge of a proton. If the charge of a proton is denoted by Q then the charge of a cobalt nucleus is $27Q$, that of the nickel nucleus is $28Q$ and it would be $13Q$ for aluminium. This was the real mystery. Mendeleev could not understand why nickel should come after cobalt, because he did not know about this. The table prepared by Mendeleev was not arranged according to the atomic weight — it had actually been arranged according to the charge of the nuclei. So now we can explain Mendeleev's discovery — even though the nuclear charge keeps increasing monotonically from element to element, the properties of the elements themselves keep returning in a periodic fashion.

Hydrogen occupies the first place in the periodic table. The charge of a hydrogen nucleus is simply Q, meaning it has only one proton. And one electron rotates around this nucleus. The charge of the electron is $-Q$, making the total charge of a hydrogen atom zero. An electron is extremely light. So the mass of the atom is more or less equal to that of the proton.

In nature, another isotope of hydrogen is found, which has an atomic weight of 2. What is the difference? The charge of the nucleus of this isotope is also Q, and it too has one electron outside. But the nucleus now contains a neutron, along with the proton. A neutron does not carry any charge. But its mass is almost equal to, in fact a little more than that of the proton. So the sum of the charge of a neutron and a proton remains Q, but the mass now becomes a little more than twice that of an ordinary hydrogen atom. But this isotope is very rare and the average atomic weight of hydrogen is therefore only slightly more than 1: in fact, 1.008, as seen from Table. 3.1.

Next comes helium, in the second place. In this case the nucleus has a charge of $2Q$ because it contains two protons. Add two neutrons to this and the mass is close to four, as can be seen from Table. 3.1. Lithium has three protons and four neutrons. This is why the atomic weight of lithium is close to 7, the charge of the nucleus is $3Q$ and it is placed in the third position in the periodic table.

Good, we have got it right. The solar system model of Rutherford describes the atoms correctly - with the nucleus in the centre and the electrons rotating around it. The number of protons in the nucleus is equal to the number of electrons outside — together they make the atom charge-less, neutral. The chemical property of an atom depends on the number of electrons. Since this number is equal to the number of protons, we could say that the chemical property is dependent on the charge content of the nucleus. The atomic weight is not important here. The charge of a lead nucleus is $82Q$, whereas its atomic weight can be 206, 207 or 208. The nucleus contains 82 protons. Each proton has a charge Q, therefore the total charge of the nucleus is $82Q$. And there would be 82 electrons outside. This is the composition of a lead atom. Now the weight of this atom would depend on the number of neutrons in it. If there are 124 neutrons with 82 protons then the atomic weight of lead would be 206, whereas 125 or 126 neutrons would give lead atoms with atomic weights of 207 or 208. Therefore, it is the difference in the number of neutrons that gives rise to different isotopes of atoms.

Now that we understand the structure of atoms, we have the option of playing God. We can actually have atoms made to order. There are 92 naturally occurring elements. Uranium is the heaviest of these — it has 92 protons in its nucleus. If we force some extra protons and neutrons in it, then newer elements can be formed. These would be completely new, which have never before been seen in nature. This was not an easy task but neither was it impossible. And this is how neptunium, with 93 protons in its nucleus, has been produced. After that about 15 new artificial elements, with 94 protons, 95 protons and so on, have been produced till now. Of course, these elements are highly radioactive and are therefore extremely short-lived. Perhaps that is why we do not see them in nature. Even if some of these elements were created in some natural process they have all decayed down to other elements long ago.

But let's keep this discussion aside. We have left the thread of history somewhere behind. Artificial elements were not created before 1940. Even the existence of neutrons in a nucleus was a mere guess in Mosley's time. Direct evidence for neutron's existence was found in 1932 — almost two decades after Mosley's discovery. Still, there is no doubt that a clear picture of the structure of atoms emerged only through Mosley's work.

3.6   Bohr

However, certain discrepancies become apparent in this picture when examined closely. Maxwell had established the theory of electromagnetic fields towards the end of the nineteenth century. According to this theory if the value or the direction of the velocity of a moving charged particle changes, then the particle radiates energy and consequently its own kinetic energy decreases. Now an electron would have to continuously change direction if it has to rotate around the nucleus. If that happens then the electrons in Rutherford's atom would continually radiate and soon would have no kinetic energy left. The less energy an electron has the closer it would go to the nucleus and finally it would be completely assimilated into the nucleus. Calculations made using Maxwell's theory showed that the time required for an electron, in an ordinary atom, to fall into the nucleus would be less than ten millionths of a second.

But the atoms are not that short-lived! We can see that everything around us is continuing as they have for years and years. Then?

And this is not the only problem. What is the energy of an electron orbiting around the nucleus? Leaving the rest mass energy aside, it would be part kinetic energy and part potential energy. According to Newton's kinetic theory the total of these two could be anything — one, two, one and a half or even 1.586329 in some units. There is a nice word to describe this situation, which we learned in Chapter 2. What we are trying to say is this — energy is *continuous* in Newtonian theory of motion.

Now let us consider firing bullets at these atoms. The electrons would collide with these bullets. As a result

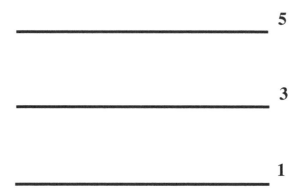

Figure 3.5: Hypothetical energy states.

sometimes the electrons would lose energy, sometimes they would gain energy from the bullets. How much energy can an electron gain in this fashion? Again Newton's law says it could be any arbitrary amount. That is, the loss and gain in energy would also be continuous.

Yet, what James Franck and Gustav Hertz found in their 1914 experiment is completely contrary to this. To understand the result of their experiment we have to look at Fig. 3.5.

Let us assume that the electrons inside an atom can only have certain fixed amounts of energy and not any arbitrary value. In other words, the energy of electrons is not continuous but discrete, as has been shown in the picture. Now let us say that an electron is sitting at the lowest level of our picture, and the energy it has is one unit. If this electron wants to move up to the next higher level it would require two units of extra energy, because the energy of the second level is three units. Our electron does not have this energy and it is therefore sitting at the lower level.

Now a bullet is fired at this electron. The bullet has only one unit of energy. Can the electron gain this one unit of energy from the bullet? No, it cannot. Because it would then have no place to go to. With extra one unit of energy it cannot sit at the lower level because now it has more energy than required. On the other hand it does not have enough energy to go to the second level. Since it cannot hang in limbo in the middle it would not take any energy from the bullet and stay put at the lower level. So the bullet would not lose any energy at all.

If instead of one unit, the bullet had one and a half or a quarter to two units of energy — we would have observed the same thing happening. The electron would have stayed back on the lower level and the bullet would not have lost any energy. But now if we increase the energy of the bullet to more than two units the situation would change drastically. The electron would become active, taking two units of energy it would go up to the next level. The bullet would lose this two units of energy and would come out very much less energetic. The situation would continue to be like this even if we increase the energy of the bullet till it becomes larger than four units. Then the electron would be able to take up four units of energy in one go and jump to the third level.

This is certainly not in accordance with Newtonian theory. The electron should have been able to gain some energy whatever may be the initial energy of the bullet. But Franck and Hertz did not see that happen.

Instead, what they saw resembled Niels Bohr's proposition. Bohr said that the electrons inside an atom can only have certain discrete values of energy. As long as an electron is in such an energy level, it would not radiate — in

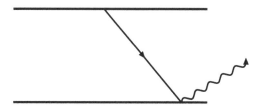

Figure 3.6: Emission of a photon by an electron descending from a higher energy level to a lower one in an atom.

violation of Maxwell's theory. Only in the case of a transition from a higher energy level to a lower energy level would an electron radiate some amount of energy, as seen in Fig. 3.6. This energy is electromagnetic in nature. According to Planck, the energy would be radiated in the form of a particle of light, the photon. Evidently, this photon cannot have any arbitrary amount of energy. If the electron comes down from the second level of Fig. 3.5 to the lowest one, then the photon emitted would have the energy equal to the difference between the energy of the levels, i.e., two units.

Same thing happens in the neon tubes used in advertising. Electricity is passed through the neon atoms, to move the electrons to the higher energy levels. When these electrons descend to their original energy levels, they emit photons which we see as light.

This light then consists of photons which have only discrete energies. According to Planck's hypothesis the photon energy is intimately related to the frequency of the light. Therefore, the light radiated would also have discrete frequencies. This was already known from analysing the spectrum of gas lights. In fact, Niels Bohr asserted that Maxwell's theory would not be applicable to the atoms, just to explain these observations.

Where is the mystery then? Almost all the propositions of classical physics have now become outdated. We have to believe the experimental results. After all, the purpose of science is to explain the observations. So, what now?

3.7 Heisenberg

In a situation like this, one needs to re-examine the most fundamental tenets of the subject. In the end, that task was undertaken in the third decade of the twentieth century.

Heisenberg said, we have missed one point. Say, an electron is moving. We would like to know its trajectory. What do we mean by this? We actually want to know the precise position and the speed with which it is moving in a particular direction at any given moment. What should we do to obtain such information?

If the discussion about electrons feels somewhat weighty we can take a break and think about a game of snooker instead. A snooker table is laid out in a dark room and only a red ball is lying on the board somewhere, its location unknown. We have the cue in hand and are allowed to use it to find the location of the red ball. Our aim is to find where the red ball is sitting.

It does not appear to be a very difficult task. Place the cue ball at some point and hit it straight against the wall opposite. If the red ball is not located along the path then the cue ball should return directly back into your hand. If, on the other hand, the red ball happens to be on the path, the cue ball would collide with it and get deflected to some other direction. It would not return back to its original position. Now we know that the position of the red ball is along the path of the cue ball movement.

Actually, this is not entirely correct. We should say that the red ball was along that line before colliding with the cue ball. But now it has moved to a different position after being deflected by the cue ball. We have changed the position of the red ball in an attempt to know its position!

The case of the electron is just like this. We are shining a beam of light on an electron to find its position, momentum, etc. If the particles of light, the photons, come back to us after colliding with the electron then only we would be able to obtain information about the electron. But we are again faced with the dilemma — whether the position or the momentum of this electron would be the same before and after the collision. The momentum of the electron would certainly change in case of an energetic collision.

Alright, let's then try for a less energetic collision. The energy of the collision would depend on the momentum of the photon — the higher the photon energy, the larger the energy of the collision. But according to Planck's hypothesis the momentum of a photon depends on the wavelength of light. If the wavelength of a given light wave is D then the momentum p of the associated photon is given by

$$p = \frac{h}{D},$$

where h is Planck's constant, which we have talked about in the last chapter.

Now this is easy. Let us take some light of large wavelength. The larger D is, the smaller p would be. If we make D very large, p would be very small. A photon with a small momentum would impart a very small amount of energy during a collision. And the electron energy would not change much as a result of the collision. We would then be able to say with some confidence that the change in the electron momentum as a result of measurement is negligible.

All that is fine. But we also want to know the position of the electron. How far have we progressed on that? A very simple rule-of-thumb is: if we try to measure the position with a wave of wavelength D, the result cannot be determined to an accuracy much better than D. It would be easier to understand this through an example. The wavelength of ordinary light is somewhere in the ballpark of one hundred thousandth part of a centimeter, or ten millionth of a meter. Because of this we need to prepare really smooth mirrors to reflect light. On the other hand, the wavelength of sound waves is a few hundred meters. That is why sound gets reflected from the the walls of a house, or from a mountain. The unevenness of the wall or of the surface of the mountain is not perceived by the sound wave. Similarly, whether an electron is located at a particular position or at a distance of $D/10$ or $D/15$ from that point would not be detected by a light wave of wavelength D. We have used a large D to know the momentum precisely, and as a result our idea about the position of the electron has become vague.

If we want to know the position of the electron precisely we would have to use a small wavelength. But then the photon momentum would be large and the collision would be highly energetic. The momentum of the electron would change significantly and we would not have any clear idea about its original momentum.

Heisenberg said that this was the reality. It is not possible to know both position and momentum precisely at the same time. If we know that the position of an object is between X and $X + x$ and its momentum is between P and $P + p$ — we know the position with an accuracy of x and momentum with an accuracy of p, then the product of x and p would have to be larger than $h/4\pi$. This fact encapsulates Heisenberg's *uncertainty principle*, the most fundamental concept of quantum mechanics.

Classical physics has no mention of this principle. Because in classical physics light does not take the form of particles but moves as a wave. There is no relation between the momentum and wavelength for such a wave and it is possible to have any combination of wavelength and momentum. But Planck has taught us about photons, we can no longer continue to use classical concepts.

Then why did we not need this till now? So far classical physics has been able to accurately describe the motion of everything — from ordinary everyday objects to heavenly bodies in the sky. There was no sign of any the uncertainty principle.

There was not, because the apparatus of our measurements were not that fine. Consider the case of the snooker ball once again. With extra fine measuring arrangements we could measure its position to an accuracy of 10^{-3} cm and momentum to 10^{-4} cm/s. If the mass of this ball is about 10 gm then the product of x and p would be about 10^{-6} units. Whereas the value of $h/4\pi$ in this unit would be close to 5×10^{-28}. Because our measuring tools are rather crude, we are going nowhere near $h/4\pi$. As a result, we can go about dealing with the problems in our everyday life without bothering about the uncertainty principle.

The mass of an electron is about 10^{-27} gm. If we want to know its position to an accuracy of 10^{-3} cm, it would be impossible to find its momentum to an accuracy or 10^{-4} cm/s. Because then the product of x and p would be 10^{-34} unit, much smaller than $h/4\pi$. Evidently, the uncertainty principle would be crucial in case of an electron. This is why it is not possible to know the trajectory of an electron accurately. Because to know the trajectory of a particle, we need to know both the position and the momentum accurately at all times. According to the uncertainty principle this is not possible. If we do not know the position and the momentum accurately at one instant of time, then we cannot find where the particle would be in the next instant, and in the next and thereafter. So we would not be able to say anything definite about its trajectory any more. Therefore, a question like, "what is the orbit of an electron rotating around the nucleus?", asked in the classical style would be quite meaningless. The new physics would not only provide new answers, it would also teach us to ask new questions.

There is no reason to be depressed by this. There is no reason to think that hereafter we would not be able to measure anything at all. For example, there would be no problem if we want to measure the position of an electron along the east-west direction and its momentum in the north-south direction. We can use a wave of exceedingly small wavelength in the east-west direction and measure the position as accurately as possible. Now we can use a wave of large wavelength along north-south to get an accurate value of the momentum in that direction. But we would run into trouble if we want to know the position as well as its momentum all in the east-west direction — as has already been mentioned.

To apply Heisenberg's principle the hydrogen atom was selected first. This was a natural choice because the structure of a hydrogen atom is the simplest. As we have said earlier — it has no neutron in the nucleus: only one proton. Outside the nucleus, there is a lone electron.

Heisenberg said it is impossible to know the orbit of this electron. We are not really bothered about that. All we want to know is the energy of this electron, also whether that energy is continuous or discrete.

Heisenberg showed that there was no way of knowing either the position or the momentum if we want to know about the energy. However, there is no problem if we want to know the angular momentum of the electron along with its energy. We would even be able to find the component of the angular momentum in a particular direction. This means that we would still be able to find a number of things about the orbit of an electron even though we would be unable to measure everything as in classical physics. We summarise this knowledge of the orbit by saying that the electron is in a particular *quantum state* — that is, it has a particular energy, a particular angular momentum, etc.

Let us go into a bit more detail. By now George Uhlenbeck and Samuel Goudsmit have discovered that an electron can have two kinds of angular momenta. It would be easier to understand through an example. The Earth is rotating around the Sun. Because of this the Earth has an angular momentum. This can be called the *orbital angular momentum*, because it originates due to the Earth's motion in its orbit around the Sun. But the Earth is also spinning around its own axis at the same time. The angular momentum acquired due to this spin is in no way connected to the Sun. The Earth would have had this angular momentum even if it were not rotating around the Sun. Similarly, an electron within an atom has two kinds of angular momenta. It has orbital angular momentum due to its orbital motion around the nucleus, and it has its intrinsic angular momentum which is known as its *spin*. According to Heisenberg, we can find the value of both of these separately. Usually these are denoted by L and S. On the other hand, it is also possible to know the components L_z and S_z along a particular direction z simultaneously. Therefore, if we know the following five quantities — energy E, L, S, L_z and S_z — then we can say that the quantum state of the electron is completely known.

The calculations made by Heisenberg on the basis of these turned out to be rather interesting. He observed that the electron inside a hydrogen atom cannot have any arbitrary amount of energy. The electron can only be in certain definite discrete energy levels — as has been seen by Franck and Hertz, as noticed in the spectrum of light, as described by Niels Bohr, as we have shown in Fig. 3.5.

Similarly, the orbital angular momentum too can only have discrete values. Even its component in any particular direction has the same characteristic. This was already suspected in the experiment performed by Otto Stern and Walter Gerlach. Same situation with the spin. But before we can understand these issues we need to check out what Schrödinger had been doing meanwhile.

3.8   Schrödinger

The dual nature of energy became known through the research of Planck, Einstein, etc., in the first decade of the twentieth century. That means in some situations energy manifests itself as waves and in some other as particles. Later in 1925, de Broglie said that the same is true of matter, it too has a dual nature. We may see it as particles or we may see it as a wave.

Schrödinger constructed an equation for such waves of matter as soon as Broglie's proposition came into existence. He even solved the equation for a hydrogen atom.

What is a wave? We have seen that if a physical quantity changes its value over time and space, it can be said to have a wave nature. Then what is the physical quantity associated with this material wave? In other words, what is it that varies over space-time to generate a material wave?

Schrödinger called this quantity ψ. This symbol is actually a letter in the Greek alphabet, pronounced something like "psi". And yes, the Greeks pronounce both the "p" and the "s", although speakers of the English language drop the "p" and pronounce it more like "sy", which rhymes with "shy".

Schrödinger was done with giving it a name. But who would explain the meaning of this? It would be Max Born, quite a few years later. He said, it is impossible to know the exact location of an electron, with a specific energy, at a given moment. But it is possible to calculate the probability of finding an electron at a given position. This probability is obtained by squaring the ψ.

We shall come to that later. But what did Schrödinger obtain by solving his equation? His results were identical with Heisenberg's. This appeared very surprising at first, because there was no similarity in their calculations. They just had the same answer. Of course, later Schrödinger showed that in reality the two calculations were very similar despite their apparent differences. It is as if they were talking in two different languages. Even though they sounded different, the meaning was the same. The conclusion arrived at by both of them is the same — energy is discrete, angular momentum is discrete and the same is true of the component of angular momentum in a particular direction.

This concept is somewhat surprising from the point of view of classical physics. Because we have not encountered discrete energy levels in classical physics. However, classical physics did have quantities other than energy taking up discrete values.

Let's consider an example. Take the string of a *tanpura*, a stringed instrument, to be one meter long. This string is tied at both ends. If we now pluck the string and release it, the string would vibrate. This vibration would make the string look like a wave.

Is it possible for this wave to have any arbitrary wavelength? Not at all. Because there is a very strong constraint. The two ends of this string are tied rigidly and therefore are not allowed to move. Fig. 3.7 shows some examples of how the string would look like if we took some photographs of it. Among the various shapes we can see a complete wave, with its crest and trough, in picture *(b)*. The wavelength is exactly one meter in this case. In picture *(a)* we can only see the half containing the crest. This means the length of the half-wave is one meter now, the complete wave would be two meters long. Then again, in picture *(c)* one and a half

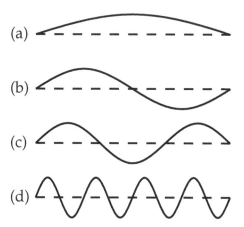

Figure 3.7: Waves in a string with two ends rigidly fixed.

wave equal one meter implying the wavelength to be 2/3 meter. Let us summarise then. The wavelength in the first three pictures of Fig. 3.7 are 2/1 (i.e., 2), 2/2 (i.e., 1), and 2/3 meter. In this way there could be wavelengths of 2/4, 2/5, 2/6 meter, etc. Like it is 2/8 or one-quarter of a meter in picture (*d*). In short, it would always be possible to find an integer n so that the wavelength could be expressed as $2/n$ meter. This means that the wavelength would be discrete — it could be two meters, or one meter but it can never be one and a half meters or one and a three quarters meter. As a result the frequency of the sound waves excited by the vibrations of this string would also be discrete. What a relief! Because the tune of a music depends on the frequency of the sound waves. If the wavelength could have taken continuous values then plucking the string of a tanpura would have excited a continuous range of frequencies at the same time. That does not happen. Usually the way the string is plucked it looks more or less like picture (*a*) of

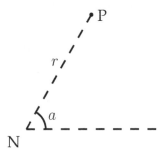

Figure 3.8: Specifying the position of the point P.

Fig. 3.7, and the sound of a particular frequency comes out. The first string of the tanpura is tied in such a way that its frequency corresponds to the fifth note *so* of the B-flat scale. That is how we hear this note. The reason behind hearing only this particular note is that the wavelength takes only discrete values. The reason behind this discreteness is the conditions imposed at the boundary — the two ends are kept fixed. And all this is happening within the purview of classical physics.

Schrödinger showed that if certain boundary conditions are imposed upon the ψ of a hydrogen atom, we would obtain a number of discrete quantities in a similar fashion. Of course, we need to justify these conditions, otherwise it would all be completely contrived. Consider the nucleus N in Fig. 3.8. The electron is orbiting around it. If we choose an arbitrary point P what would be the value of ψ? That can be obtained from Schrödinger equation if the position of P is given. We need to know three quantities in order to know the location of P. Two of these have been shown in the picture — r, the distance between N and P, and the angle a

that NP makes with a fixed direction. If we now also say that the point is at a height z from the plane of the paper, then the position gets completely determined. Of course, the value of z is zero for the P shown in the picture since it is lying on the plane of the paper. Increasing the value of z would make it rise above the plane of the paper. Similarly, an increase in the value of r would take the point further away from the nucleus. But what happens if a is increased? Going from zero to $360°$ completes one full circle. If a increases further the same points would keep coming back, because $a = 10°$ and $a = 370°$ are equivalent. So if r and z are kept fixed ψ's at $a = 10°$ and at $a = 370°$ should be equal, because they are being measured at the same point actually. Alternatively, we can say that ψ should have the same value for $a = 0°$ and $a = 360°$.

If we impose this condition on Schrödinger solution of ψ then the discreteness of L_z comes out automatically. It is seen that the value of the z-component of angular momentum is $m_l h / 2\pi$ where m_l is zero or an integer. Meaning, the component can be zero or $h/2\pi$ but nothing in between. On the other hand, it could also be $-h/2\pi$, i.e., in the negative z-direction, beneath the plane of the paper. Similarly, it could be $2h/2\pi$, or $3h/2\pi$ or $-5h/2\pi$ — but it can never take any arbitrary value.

There is something else. We have already said that the probability of finding the electron at any point is obtained by taking the square of ψ at that point. Now probability also cannot take any arbitrary value. If an event is totally improbable we would say that its probability is zero. On the other hand if an event is absolutely certain to take place its probability would be 1. If we are not certain of either of these two cases then we say that the probability of this event is between 0 and 1. This means that probability is never smaller than 0 and is never greater than 1. This is another justified condition that should be imposed upon the ψ of an electron.

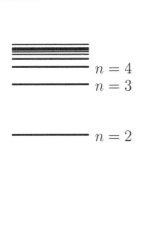

If this condition is imposed at the location of the nucleus then we find that the value of L^2 can also be written in the form $l(l + 1)(h/2\pi)^2$. Again this l can only be zero or some positive integer. Meaning the value of L^2 could be zero (if l is zero), or twice that of $(h/2\pi)^2$ (if $l = 1$), or six times (if $l = 2$) or twelve times ($l = 3$) and so on. On the other hand, if we say that the probability of finding the electron very far from the nucleus would have to be small we would be able to arrive at the conclusion that the energy of the electron is discrete. The allowed energy values could be written as,

$$E = -\frac{2\pi^2 m_e Q^4}{n^2 h^2}.$$

Figure 3.9: Electronic energy levels in a hydrogen atom.

Moreover, n in this equation is a non-zero integer. h is our familiar Planck's constant, m_e is the mass and Q is the charge of an electron.

We shall obtain different energy levels for different values of n ($n = 1, 2, 3, 4, 5\ldots$) as has been shown in Fig. 3.9. Let us here assure those who are getting confused by seeing a negative expression for the energy. This is nothing un-

n	any integer starting from 1
l	any integer from 0 to $n-1$
m_l	any integer from $-l$ to l
s	$\frac{1}{2}$
m_s	$-\frac{1}{2}$ or $\frac{1}{2}$

Table 3.2: Ranges of the atomic quantum numbers.

usual, just a way of writing. After all the electron is bound to the nucleus, not escaping away. Why can't it do that? Because it does not have enough energy. Now if we say that it has an energy equal to -13.6 units, it means that if it had an extra 13.6 units of energy it could have escaped.

This is how we obtain the value of energy. But knowledge about energy is not sufficient to have complete information about a quantum state. We also need to know the values of L, L_z, S, S_z. The value of L depends on l, that of L_z on m_l. Schrödinger showed that according to his equation l has to be smaller than n. That is, for a given value of n, l can take up values from 0 to $n-1$, but not more. For example, if n is 3 then l can be 0, 1 and 2. Similarly, m_l can start from $-l$ and can increase up to l. If we now write the value of S^2 as $s(s+1)(h/2\pi)^2$ by comparing it to L^2, then for an electron s would always be 1/2. And S_z could only take two values — $-h/4\pi$ or $h/4\pi$. Of course, this last part has not come from Schrödinger equation. It came from some other considerations which we would discuss at a later point.

We have amassed a lot of information in this section, and it may be difficult to remember all of it. So let's go through this once again then. The basic fact is that we need to have five quantities, n, l, m_l, s and m_s, in order to know about a particular quantum state of an electron. Amongst these s

is always 1/2, the rest can have different values. To help us remember this we have summarised the range of values taken up by these five quantities in Table. 3.2.

3.9 Pauli

What is the situation then? When $n = 1$ how many quantum states of electron are there? If n is 1 then l has to be zero, because l has to be smaller than n. Moreover, l can never be negative. If l is zero then m_l would also be zero. s is always 1/2. m_s can, of course, be 1/2 or $-1/2$. Altogether, we have two quantum states then, one with $n = 1, l = 0, m_l = 0, m_s = 1/2$ and another with $n = 1, l = 0, m_l = 0, m_s = -1/2$.

What if we have $n = 2$? Now l could be 0 or 1. When $l = 0$ we have two quantum states like before. But if $l = 1$ then we have three possible values for m_l viz., -1, 0, 1. For each of these cases we shall have m_s equal to either 1/2 or -1/2. In all, we have six quantum states. If we add the two states coming from $l = 0$ case, the total number of quantum states would be eight for $n = 2$.

Now Wolfgang Pauli prescribed that there cannot be more than one electron in a quantum state. The fancy name for this prescription is the *exclusion principle*. That is if there is one electron in a given quantum state, it would be excluded by other electrons.

Hydrogen has only one electron. Under normal conditions the value of n would be the lowest, i.e., 1. Then both l and m_l would be zero and m_s can be either +1/2 or -1/2. Helium has two electrons. If one of them has $n = 1, m_s = 1/2$ then the other cannot have the same values.

The second electron would have to have $n = 1, m_s = -1/2$. Well done. It is now full house for $n = 1$. If we have another electron it cannot be accommodated at the $n = 1$ level. This shell has become fully occupied. Elements with this type of filled shells are inert. The first example of an inert element is helium. We shall soon learn about other such inert elements.

Lithium. It has three electrons. Two of them would go to the $n = 1$ shell, just like in helium. Where would the third one go? It cannot go to $n = 1$ shell, by Pauli's exclusion principle. So it has to go to $n = 2$ level. So lithium has an inert core like helium, and a single electron outside like hydrogen. Is there any difference from hydrogen? Of course, hydrogen does not have a core. But that core is inert. So if we ignore that inert part, lithium has the same electronic structure as hydrogen. That is why lithium is so similar to hydrogen, and is sitting in the same column as hydrogen in Mendeleev's periodic table.

Next comes beryllium. This too has a core like helium. But now there are two electrons outside. However, it would be wrong to quickly club it with helium simply because it has two electrons outside. Because even though it has two electrons outside, the $n = 2$ shell has not been filled up. We have seen that the total number of quantum states corresponding to $n = 2$ is eight. So this is something new. Beryllium has two electrons outside, apart from its inert core. Hydrogen and lithium had one electron each on the outside and helium had none. So the chemical property of beryllium is different from all of these. A new column has to be created for beryllium. That is exactly what Mendeleev did. Similarly, boron has three electrons apart from the inert core, carbon has four, and finally fluorine has seven electrons on the outer shell. Then we reach neon. Now there are eight electrons in the $n = 2$ shell, plus the two in the $n = 1$ shell. The two shells are now completely filled with ten electrons making neon the second inert element in the periodic table occupying the tenth place.

Sodium has eleven electrons. Ten of them are in $n = 1$ and $n = 2$ shells and the final one in the $n = 3$ shell. Once again the situation is similar to hydrogen and lithium — one electron outside apart from the inert core. Naturally they have similar chemical properties. Magnesium has two electrons in the outer shell and is similar to beryllium. Then we have aluminium which is similar to boron, silicon to carbon, phosphorus to nitrogen and it would continue like this.

Now we can see the underlying meaning of the periodic table. If we ignore the inert cores then the elements with same number of electrons in the outermost shell would have similar chemical properties. We would have to use Pauli's exclusion principle to find the elements that would have the same number of electrons in their outer shell. We have already understood the first few rows of the periodic table using this rule. Later on the solutions of Schrödinger equation would become very complicated. We can explain the entire periodic table using Pauli's principle if we brave those complications. However, now that we know the fundamental principles we can skip the mathematical details and leave things at that.

3.10   In short

We have now gained a fair understanding of the interior of an atom. Everything on Earth is made up of atoms, and there are ninety-two types of naturally occurring atoms. We know that an atom is not indivisible. There exist three types of particles inside an atom — the protons and neutrons combine to form the nucleus and the electrons orbit outside it. The chemical property of an element depends on its number of electrons. Because the number of electrons is equal to the number of protons, we can also say that the chemical property depends on the number of protons. And a difference in the number of neutrons in the nucleus gives rise to different isotopes. We have also learnt about the structure of different atoms, and the similarity of their chemical properties.

We have understood the notions of discreteness of energy, the limitations of classical physics, the beginning of quantum mechanics and its development. Interestingly, the key to the chemical mysteries also lay hidden in this solution to the most fundamental crises of physics. In a sense, the summary of this chapter is actually the story of chemistry merging with physics. These two subjects are not really separate, simply different manifestations of the same truth.

And the dream of science is just that — to find connections between completely disconnected events, to combine apparently disparate ideas. We have talked about such synthesis in previous chapters, continued in the same vein here and would take it further ahead in the later chapters.

Chapter 4

Smaller than the atom

4.1 Preamble

The English word "atom" has been derived from the Greek word "tomos", which means "body". Thus "a-tomos" means something that does not have a body, meaning something that does not have any component part. In other words, the object is indivisible into smaller components, or parts. We have already learned about the interior of an atom and know that there is nothing "indivisible" about it. An atom is actually made up of a number of smaller particles. In this chapter we shall deal with such particles, particles that are smaller than an atom.

We have, in fact, talked about some of these sub-atomic particles already. We know that the nucleus at the centre of an atom contains protons and neutrons, and a group of electrons orbit around them. Atoms of every element are made up of this three types of particles. Earlier we have also discussed the concept of particles of light, or photons, which are necessary for understanding the true nature of light or electromagnetic fields in general.

This might give one an impression that the elementary particles constitute a small four-member cozy family. Anything more complex can be formed from a combination of these. Planck and Einstein explained the behaviour of photons. And the nature of atoms can be obtained from quantum mechanics after Schrödinger and Heisenberg. Of course, quantum mechanics was still in its infancy but already physicists could see its enormous potential.

Not only the lay public, even physicists would have thought by the end of the third decade of the twentieth century that the key to all the mysteries of the universe was within their grip. The only task that remained was to solve the Schrödinger equation for various difficult situations. It was fortunate that before any such thought made the physicists content, a number of important events took place one after another within a span of few years. Because of these it became evident that the world of elementary "particles" is not so simple.

4.2 Dirac

Einstein's theory of relativity came into existence in 1905. It showed that Newton's theory of motion does not contain the ultimate truth but has limitations. If the velocity of an object becomes very large, comparable to the velocity of light, the theory of relativity would be necessary to describe its motion.

Similarly, with the advent of quantum mechanics in 1925 we saw that in the world of microscopic particles — in the interior of an atom — classical physics is not adequate.

Newtonian theory of motion is then running into trouble from both directions — in the high velocity regime, as well as in microscopic dimensions.

What happens if both of these conditions are realised together? For example, consider an electron within an atom having a velocity comparable to that of light. What would happen then?

In such a situation we require a theory known as *Relativistic Quantum Mechanics*. This theory tells us how to explain the behaviour of high-velocity sub-atomic particles, by combining relativity with quantum mechanics.

Paul Dirac brought about this synthesis in an article published in 1929. But the importance of Dirac's theory is not confined to just this.

Take the spin of an electron, for example. We have said that the spin of an electron is actually its internal angular momentum. To explain this we gave the analogy of the Earth. We said that the spin of an electron is similar to the rotation of the Earth about its own axis.

Let us look at this comparison in some detail now. Did we mean that the electron is also a sphere like the Earth, only extremely small — and a rotation around its own axis is giving rise to its spin?

Initially, this was exactly the kind of picture people had in mind. The word 'spin' also came from such considerations — assuming a very small sphere of radius $\sim 10^{-12}$ cm to rotate like a top. This rotation is giving rise to the spin.

But there is a problem with this line of thought. The spin of an electron is always $\frac{1}{2}$, that is, the value of the square of its spin angular momentum is always $\frac{1}{2}(\frac{1}{2}+1)$ in units of $(h/2\pi)^2$. Irrespective of the state of the electron its spin would always be $\frac{1}{2}$. If this spin is due to the rotation of a small sphere how could it be completely invariant? We can always reduce the angular momentum of a top by touching it. Friction with air molecules can also reduce angular momentum. Even the huge Earth is losing its angular momentum, of course, extremely slowly — but the angular momentum is decreasing. Why does this not happen in case of an electron?

The first victory of Dirac theory was in this. When Dirac set up the equation for an electron according to the relativistic quantum theory, it was seen that there could only be four states of a spin-half particle. It is very easy to see how two amongst these four states arise. We have already seen that there can be two components of spin (or m_s) given by either $+\frac{1}{2}$ or $-\frac{1}{2}$ for an electron which is a spin-half particle. These two states, i.e., states with two different components of spin, are obtained from Dirac equation. And to obtain these, there is no need to think of the electron as a small sphere, or that it is rotating about its axis. Even if the electron had zero velocity it would have this angular momentum. Therefore, it is the *intrinsic* angular momentum of an electron, which does not depend of anything external and consequently is totally invariant.

If this looks surprising then it needs to be mentioned that similar situations have been encountered before. But we did not really admit it. For example, consider the rest energy of an object. The kinetic energy comes from the motion of an object: that is easy to understand. But the rest energy? Where does it come from? To avoid this question we have said that the energy associated with the rest mass is the rest energy. Where this rest mass is coming from then? We do not know. If pressed we simply say that the rest mass is an intrinsic property of an object. The rest energy is its intrinsic energy. Why does an electron have charge? We do not know. We say that it is an intrinsic property of an electron. Similarly, spin is intrinsic. It has no relation to any rotation or any motion at all.

There is nothing very unusual in this, so far. Dirac simply explained some of the known concepts clearly. Any good theory does that. But it also makes new predictions. Dirac theory was no exception to this.

Dirac calculated the energy levels of the electron in a hydrogen atom using his new equation. It was observed that the small mismatches that the experimental results had with the Schrödinger solution have now gone away in this new theory. The second success of Dirac.

Now we have to face a difficult question. We have said that Dirac found four states of an electron from his equation. We understood two of them already. But what about the other two? Where do they come from? What is their significance?

After a lot of thought Dirac realised that this was actually another type of particle. It too has a spin of $\frac{1}{2}$ like an electron, and can have two possible values of components of spin, $+\frac{1}{2}$ or $-\frac{1}{2}$. This particle has the same mass as the electron. What is the difference then? The charge of this particle is exactly opposite to that of an electron. The charge of an electron is negative, and it is positive for this new particle. Though the magnitude of the charge is equal for both of these. Dirac called this other particle the *anti-particle* of electron. It has a very intimate relation with electron — if electron exists this anti-particle would also have to exist.

This is why the same equation generates the quantum states corresponding to both the particles.

That brings in a new problem. We know about only one type of positively charged particles — the protons. The spin of a proton is also the same as that of an electron, $\frac{1}{2}$. But mass? The mass of a proton is 1840 times that of an electron! Dirac was in a quandary. And said that he must have made a mistake in the calculations. Perhaps the mass of an anti-particle need not be the same as that of the particle. Proton has to be the anti-particle of the electron — because there exists no other particle with a positive charge.

Alas, Dirac was defeated by his own results. There was no mistake in his calculations. In 1932 a new particle was discovered in Carl Anderson's instrument. It had the same charge and spin like the proton, except for the mass. The mass, in fact, was equal to the mass of an electron. This is the particle that Dirac found through his equations, only did not have the courage to accept its existence. Because of the positive charge, this particle was dubbed as *positron*.

From where did the positron arrive in Anderson's instrument? The outer space. Uncountable number of energetic processes are taking place in the infinite space outside of Earth — the stars are burning bright, they are colliding with each other and so on. In each process, particles are being created and/or destroyed. Most of these particles, collectively known as *cosmic rays*, are moving through space at extremely high speed. Sometimes they run aground on Earth. Anderson found his positron in the cosmic ray.

Why is the positron absent amongst the constituent particles of Earthly material? Why do we only have electrons, protons and neutrons here? Because, we must remember that the positron is the anti-particle of electron. The mo-

ment a positron meets an electron there would be total annihilation of both the particles and we shall be left with only
photons, i.e., electromagnetic energy. Evidently, a positron
coming from the outer space would be destroyed upon its
encounter with an electron, since the Earth is full of electrons. Therefore, we need to catch a positron before the
annihilation took place. Anderson was successful in doing
just that.

Is it possible to artificially reproduce some of these reactions, taking place in outer space, on Earth? Those require huge amounts of energy. Such energies are not completely out of reach of modern instruments like cyclotrons,
betatrons, etc. So we can create positrons on Earth too.
Though the life of such positrons would also be very small,
we would not have to wait for them to arrive from the outer
space.

And it is not the question of positrons only. According
to Dirac theory we are supposed to have the anti-particles
of every particle known to us. These anti-particles would
have the same mass and spin corresponding to their particle, only the charge would be of opposite sign.

In reality, it does not make sense to call something a particle or an anti-particle. Instead the two particles together
should be called a particle and anti-particle pair, like electron and positron. Similarly, the proton and antiproton or
neutron and antineutron are such pairs. And none of these
particles are products of fantasy. The antiproton was seen
by Emilio Segre, Owen Chamberlain, and their associates in
1955.

Our list of the elementary particles, therefore, is by no
means complete. The first definite indication for this came
from Dirac's theory and Anderson's experiment.

4.3   Pauli

We talked about the discovery of neutrons at the end of the first chapter. Some of the radioactive elements emit an electron when they decay down to other elements. This is known as *β-radioactivity* using the second letter of the Greek alphabet, β (beta).

It needs to be remembered that the nucleus of an atom does not contain any electron. Then how would an electron be emitted? The answer to this is that such electrons are produced, along with a proton, by the decay of a neutron. We write this like a chemical reaction as follows,

$$n \to p + e$$

that is,

$$\text{neutron} \to \text{proton} + \text{electron}.$$

Say the original nucleus, X, had Z protons and N neutrons. This can be written in short as X_N^Z. If one of these N neutrons decay and give rise to one proton, then finally there would be $(N-1)$ neutrons and $(Z+1)$ protons. If we call this new nucleus Y then we could write this reaction as,

$$X_N^Z \to Y_{N-1}^{Z+1} + e.$$

The charge of an electron is equal and opposite to that of a proton. So the total charge of an electron and that of $(Z+1)$ protons is still equal to that of Z protons. The electrical charge is therefore conserved, without any problem.

However we have learned about many other conservation laws, besides that of the electric charge. For example, the momentum. If initially the nucleus, X, was at rest then the sum of the momenta of the ejected electron and the nucleus, Y, should also be zero. So the momentum of the electron should be equal and opposite to that of Y. But that was not what the experiments indicated.

There were problems with the conservation of energy too. If the rest mass of X and Y are denoted by M_X and M_Y then their rest energy would be $M_X c^2$ and $M_Y c^2$. Similarly, the electron has a rest mass m_e and rest energy $m_e c^2$. What about the kinetic energy of the particles? X was at rest and therefore had zero kinetic energy. We can also ignore the kinetic energy of Y as it would be insignificant. Then, if K_e is the kinetic energy of the electrons, according to the law of conservation of energy we should have,

$$M_X c^2 = M_Y c^2 + m_e c^2 + K_e,$$

or

$$K_e = (M_X - M_Y - m_e) \times c^2.$$

That is, the kinetic energy of the electron should only depend on the rest masses of various particles. Since rest mass of a particle is a constant, the value of K_e should also be a constant. This means that if we look at a million nuclei, and if a million electrons are emitted because of their radioactivity, then all of them should have the same kinetic energy. But James Chadwick showed in 1914 that such is not the case. The kinetic energy of the electron could be anything between zero and $(M_X - M_Y - m_e) \times c^2$. Similarly, there were problems with the conservation of angular momentum.

Niels Bohr and company then proclaimed that there was no need to worry. Since these conservation laws belonged to the realm of classical physics, they were not relevant for quantum mechanics. We already know that classical physics cannot explain everything, so there is no reason to be surprised by this.

But Wolfgang Pauli was not ready to concede defeat without a fight. In December of 1929, Pauli sent a letter to the scientific convention taking place in the city of Tübingen in Germany. This letter basically said that the conservation laws are correct, it is the scientists who are making a mistake. The fact is, in the decay of a nucleus another particle is also emitted along with the electron. Today we call this particle an *electron-antineutrino* denoted by $\bar{\nu}_e$. The Greek symbol ν corresponds to a neutrino. The subscript e means that it is emitted along with an electron. And the bar on top is to say that it is an antiparticle. In the same manner we denote positron by $\bar{e}$ and antiproton by $\bar{p}$.

In other words our version of the basic equation for radioactivity itself was wrong. It should have been,

$$X_N^Z \rightarrow Y_{N-1}^{Z+1} + e + \bar{\nu}_e \,.$$

It is clear that this new particle is charge-less, otherwise we cannot have charge conservation. Till then the methods used to detect radioactivity could only detect charged particles. That is why neutrino was never seen. If the mass and momentum of the emitted neutrino were measured there would not have been any problem with the conservation of these quantities. Similarly, this particle also has spin, the value of which is equal to that of the electron. The conservation of angular momentum was in trouble because this spin was not accounted for.

Still mere conjecture is not enough, direct proof was needed for this. Though it was soon realised that the task would not be an easy one. Assuming the existence of neutrinos, the equations to describe radioactivity were set up by Enrico Fermi. From these it was seen that the probability of collision between a neutrino and any other particles is extremely small. How then could a neutrino be detected? In order to detect a particle we need the particle to collide with the material of the detector used. If there is no collision then the particle would simply pass through the material of the detector without making its passage known.

The collision probability of a neutrino is so small that initially it was thought to be quite impossible to detect. Then around the 1950s atomic reactors started to be built for the purpose of energy generation. The atomic reactions taking place in these reactors were producing copious amounts of neutrinos. The number of the neutrinos was so huge that if even a minuscule fraction of that were detected it would generate a clear signal. Using this particular technique the neutrino was detected by Frederick Reines, Clyde Cowan, and their associates in 1956, i.e., about twenty-seven years after Pauli's announcement.

Still we have to concede that the scientists more or less accepted the existence of neutrino following Pauli's conjecture. It is evident from Dirac's theory that this particle would have its own antiparticle too. There is no prize in guessing that the name of the antiparticle is *electron-neutrino* and it is denoted by the symbol ν_e.

4.4 Yukawa

The nucleus of an atom has a lot of secrets. To begin with, consider radioactivity. Why does a nucleus decay spontaneously? We shall come to that in the next chapter. For the moment we need to resolve a more serious problem regarding the nucleus. Isn't the existence of the nucleus a puzzle in itself?

Some people might say that the *existence* of anything is a big puzzle. No, we are not dealing with such philosophical questions here. There is a reason for our worry though. The nucleus of all atoms, except that of hydrogen, contains more than one proton and neutron. The neutrons do not have any charge and therefore are immune to any electrical force. But the protons are positively charged. We know that any two positively charged particles would repel each other. Then how could the protons stay put inside the nucleus? Why are they not being blown away due to this force of repulsion?

At first we might think that there is no reason to be overly concerned. Because there also exists the force of gravitation between any two particles, which is always attractive. If this attraction between two protons happens to be larger than the force of repulsion then the resultant effect would be that of attraction. And the protons could reside together inside the nucleus. The neutrons too would be attached to the nucleus due to this attraction.

Sadly, that is not a viable explanation. A back of the envelope calculation shows that the gravitational force compared to the electrical force between two protons is not merely small, but completely insignificant — about 1 part in 10^{39}. So we are back to the same question. What is it that keep the protons bound to the nucleus?

There could be only one answer. There must be another force acting between the protons, besides the gravitational and the electrical. The value of that attractive force is larger than the repulsive electric force. This force then ensures that the protons remain bound to the nucleus. This was given the name of *strong force*.

We shall talk about strong force later. First we need to clarify something else. Till now we did not need to invoke this strong force! We have calculated the energy levels of the electrons inside an atom without talking about strong force. And the results were not incorrect. Why?

Of course, there is a reason for that. Fact of the matter is, the electrons are not responsive to this strong force. Just like the neutrons or neutrinos are not responsive to electric force because of their lack of charge. A proton is a charged particle and is therefore responsive to electric force. A proton is also responsive to the strong force. All particles that respond to the strong force are known as *hadrons*. For example, the neutron — oblivious of the electric force, but behaves quite like the protons under strong force. So neutron is a hadron, same as a proton. But the electron is not a hadron, neither is a neutrino. So we can forget the strong force in the context of such particles.

All that is fine. But if the strong force is really this strong then that should also make all the nearby nuclei merge together. Why does that not happen either?

This is not happening because under ordinary, everyday conditions the average distance between two nuclei of any material is about 10^{-8} cm. The strong force is very powerful, but it also has a very short range. A distance of 10^{-8} cm is rather huge for this force. We know from Rutherford's famous experiment that a nucleus is an extremely small object, with radius around 10^{-13} cm. At such small distances the strong force is most effective but if the distance scales are about 10^{-8} cm then the effect of strong force becomes quite insignificant. This is why protons and neutrons residing only within the range of force field around a proton, which is about 10^{-13} cm, feel the attractive strong force.

There is no effect outside of this range, a nucleus at a distance of 10^{-8} cm does not even know about this force field.

We have actually mentioned something rather important in the course of this discussion. The way the concept of electromagnetic field arises in the context of the electromagnetic force, we have to talk about the *strong field* in the context of strong force.

But we have said that to explain some of the characteristics of the electromagnetic field we need to invoke a kind of particle — the photons. Then what would be the corresponding particle for the strong force?

It is a new particle — said Hideki Yukawa. His calculations showed that the rest energy of this particle would be between 135 and 140 *MeV*. Those who are not familiar with MeV need only remember that this is a unit of energy. Measured in grams the rest mass of an electron is 9×10^{-28} and that of the proton is 1.6×10^{-24}. In order to avoid dealing with these minuscule numbers we are choosing a unit such that the numbers look somewhat more respectable in it. MeV is one such small unit. In this unit the rest energy of electron is a little larger than 0.5, and that of the proton is 938. This means that the new particle proposed by Yukawa is almost 275 times heavier than an electron. He found that this is a spin-zero particle. Yukawa also found the nature of the interaction of this particle with protons and neutrons.

Today we call this particle of Yukawa by the name of *pion*. Experimentalists started looking for this particle. They did find it eventually, but only after some complication. We shall talk about that now.

4.5   Rabi's question

In 1938, Carl Anderson and Seth Neddermeyer found another new particle while looking for Yukawa's pion. The rest energy is close to 105 MeV, i.e., almost 210 times that of the electron. Initially it was felt that this itself was the pion. Perhaps Yukawa's calculations were not quite precise.

Soon this idea had to be given up. Because it was seen that the properties of this particle are completely different from those of the pion proposed by Yukawa. On the other hand this was very similar to an electron with same charge and same spin. There is absolutely no difference between the behaviour of an electron and this particle in every possible situation. The new particle is also insensitive to the strong force and therefore is not a hadron. Because of the equality of charge it behaves in a similar manner to an electron in an electromagnetic field. The only difference is in the rest mass, as we have already stated. Except for the mass it is like a 'xerox' copy of the electron. This particle was named *muon* after the Greek letter *mu*, just like pion has been named after the Greek letter *pi*.

Now, electron has a 'xerox' copy. What about the electron-neutrino? In 1962, it was shown by Jack Steinberger, Leon Lederman, and Melvin Schwarz that indeed the muon too has a companion — the *mu-neutrino*. More than a decade after this, in 1975, Martin Perl and his associates discovered another particle. This has been given the name *tau* after another letter in the Greek alphabet. A tau is almost seventeen times as heavy as a muon, but there are no other differences between this particle and a muon or an electron. In fact, this too has its own accompaniment — the *tau-neutrino*. These six particles — electron, muon, tau and their neutrinos — are known as the *leptons*. Moreover each of these have their own anti-particle, i.e., six anti-leptons.

There is a story called *Bharatvarsha* (The India) by S. Wazed Ali. It is a narrative by the author in which he talks about a grocer's shop. In this shop he sees an old man reading the *Ramayana* to his grandchildren, and the young son busily going about the running of the shop. He comes back to this place again twenty-five years later. And he encounters the same scene in that shop — an old man reading the *Ramayana* to his grandchildren and the young son busy with the shop. Everything continues in the same manner, in the same tradition — only the generations have changed along the way.

It is a similar experience when we hear the story of the leptons. First the electron, and the electron-neutrino — these were investigated in detail, properties noted down carefully. And then, the same thing happens again. Another pair, exactly like before — the muon and the mu-neutrino. Like we are looking at the next generation, just like the story above. And then comes the third generation — the tau and the tau-neutrino.

Everything in nature happens for a reason. We believe that there is nothing arbitrary about nature. We see that to make an atom the required particles are — electron, proton and neutron. Similarly, photon is needed for the electromagnetic field and pion for the strong field. We can even understand the justification for having the electron-neutrino in order to conserve energy, momentum, etc., in radioactive reactions. But what purpose do the muons serve? The famous question "Who ordered that?" by Isidor Rabi voiced this collective surprise at the discovery of the muon.

Who ordered the muon, who wanted the second or the third generation of leptons?

And if there exist second and third generations what could be the reason for them to behave exactly like the first one? Why should they copy those behaviours blindly? We never felt that the nature is lacking in variety. Looking at these particles we cannot help wondering what could be the reason for producing almost the same particles over and over again? We shall face this question again in many different contexts. But the answer to this is yet to be found by the scientists.

Superficially, of course, one can think of a number of dissimilarities between the particles of different generations. For example, the electron is a stable particle. If an electron is left alone, it does not decay.

It is not difficult to see why that does not happen. In the decay of an electron, particles which are lighter than electrons would have to be produced! There do exist some particles which are lighter than an electron — like the photon or the neutrino. But all of these have zero charge. Where would the charge of an electron go after the decay then? The charge cannot simply vanish since we know that the electric charge is conserved. It stands to reason then that the electron is stable, has an infinite life, due to this conservation law.

The situation is different in the case of a muon. All the particles lighter than a muon are not charge neutral. The electron itself is lighter than the muon. Therefore, the muon is not absolutely stable, it can decay. The easiest way to explain this is to write an equation like that for a chemical reaction:

muon $\rightarrow$ electron + electron-antineutrino + mu-neutrino , or, we can write it in symbols in the following way,

$$\mu^- \to e^- + \bar{\nu}_e + \nu_\mu .$$

That is, the decay of the muon produces an electron, an electron-antineutrino and a mu-neutrino. The time taken by this process is about 10^{-6} s, or one millionth of a second. This is why it is difficult to find a muon, this is why it is not a constituent part of any atom. The muon too was discovered for the first time in cosmic rays, just like positrons. If a muon is created artificially it would decay over a timescale of 10^{-6} s.

The same thing would happen with the tau. And it can happen through either of the two following channels – either

tau $\to$ electron + electron-antineutrino + tau-neutrino ,

or

tau $\to$ muon + muon-antineutrino + tau-neutrino .

The decay of a tau through any of these channels would take about 10^{-12} seconds. The muon in the second equation would decay again. And in the end we shall have an electron.

This means that the muon and the tau are unstable particles, whereas the electron is completely stable. This is a major difference in their characteristic properties. Still we need to remember that even though this difference looks significant, the underlying reason is very simple — the rest mass of the tau or the muon is larger than that of the electron. We shall see that all apparent differences between these three generations of particles actually arise from this one difference. Due to this large rest mass of the muon, the decay of a radioactive nucleus always produces an electron and electron-antineutrino pair, never a muon and a muon-antineutrino. Because the energy released by a nucleus is smaller than even the rest mass of a muon. This energy cannot possibly give us another muon, even less a tau.

But there is no dearth of energy when a tau decays. Consequently, the probability of producing an electron or a muon are the same. The equality of these two probabilities itself shows the similarity between the properties of the electrons and the muons.

Now, we wonder whether the muon could decay in some other fashion. Let us consider the reaction given below,

$$\text{muon} \to \text{electron} + \text{photon}.$$

A photon does not carry any charge, therefore there is no problem with the conservation of charge. In principle there could be no problem with the conservation of energy, momentum or angular momentum either. Yet, a reaction like this has never been observed. Why?

Scientists drew our attention to the details of the reactions depicting the decay of a muon or a tau. Several quantities are being conserved here. For example, take the electron number. Now wait. What exactly is that? Add the total number of electrons and electron-neutrino and subtract from this the total number of positrons and electron-antineutrino. The result gives us the electron number. In other words, the electron number of an electron and its neutrino is 1, whereas that of the positron and and an electron-antineutrino is -1. All other particles have zero electron number. Now we can calculate the total electron number without any problem. If we calculate this number before and after a given reaction, we would see that the electron number does not change. The same would also be true of the muon number and the tau number. We have not really defined these two numbers, but it should be easy to understand if we look at Table. 4.1. Of course, the word *lepton number* has been used in this table for the first time. But it is obvious that the lepton number of any particle is the sum of its electron number, muon number, and tau number. This is why all the leptons have lepton number 1 and all their antiparticles have lepton number -1. The rest of the particles have zero lepton number, which has not been shown in this table.

Now let us consider the decay of a muon. We start with the muon number being equal to 1 because before the decay, we have the muon and nothing else. The final muon number is also 1 since then we have the mu-neutrino. The electron number is zero in the beginning. The final product contains an electron with electron number equal to 1, but it is accompanied by an electron-antineutrino with electron number -1. Therefore the total electron number is again zero. The same situation would be seen in the equation of tau decay. The tau number is 1, both at the beginning and at the end, whereas both the electron number and the muon number remain zero. On the contrary, if the decay of muon produced only an electron and a photon then the muon number would have been 1 in the beginning but zero in the end. In this situation, the muon number would not have been conserved and the case of the electron number would have been the same. This is the reason why we never see such reactions taking place in nature.

This is a novel kind of chemistry. In ordinary chemical reactions the reactants are the atoms, whereas here the ac-

Particle	electron number	muon number	tau number	lepton number
e-neutrino	1	0	0	1
electron	1	0	0	1
mu-neutrino	0	1	0	1
muon	0	1	0	1
tau-neutrino	0	0	1	1
tau	0	0	1	1
e-antineutrino	-1	0	0	-1
positron	-1	0	0	-1
mu-antineutrino	0	-1	0	-1
antimuon	0	-1	0	-1
tau-antineutrino	0	0	-1	-1
antitau	0	0	-1	-1

Table 4.1: The leptons and their lepton numbers.

tors are subatomic particles. The chemical mysteries were unveiled for the first time by Dalton's theory. There we learned that the number of atoms do not change in a chemical reaction, it remains conserved. Similarly, we find newer and newer conservation laws. The conservation of the electron number, the muon number or the tau number are but a few examples of that.

4.6 Lawrence's legacy

A little after the muon, the pion was discovered. The properties of this particle matched Yukawa's forecast quite accurately. Even then it soon became clear that understanding the nature of hadrons would not be as simple as it was for the leptons.

Let us look at the first complication then. All leptons are spin-half particles, and are totally alike in that respect. But the hadrons? Protons and neutrons — known as nucleons because they are found in the nucleus of atoms — have half-spin. But a pion does not have any spin, it is a spin-zero particle. The nucleons are of two types — the protons and the neutrons. Similarly, the pions are of three types. One of these is positively charged, another negatively — the value of the charge being equal to that of a proton for both. The third pion is neutral, it does not carry any electrical charge.

This means that there are two types of hadrons. The modern terminology for the ones with zero or integral spin is *meson*. And those with spin equal to $\frac{1}{2}$, $\frac{3}{2}$, etc., are known as *baryons*. Therefore the nucleons are baryons and the pions are mesons.

But the list does not end here. By now a number of accelerators have been invented, which accelerate particles to extremely high speed. These accelerators can be of different types. Ernest Lawrence thought up the first accelerator in 1930s, and the cyclotron was built. The enormous machine that has been built in Salt Lake near Kolkata is simply a sophisticated version of this cyclotron. There are accelerators in Europe (near Geneva) and in America (near Chicago) which can impart kinetic energy to a proton that is more than a thousand times its rest energy. Let us recall that the kinetic energy of each proton in a rocket moving with a speed of 25,000 kilometers per hour is less than one part in two billion of its rest energy. This may give us a clue as to the huge amount of energy that can be controlled by modern scientists.

Of course, the accelerators were not so advanced in the 1950s or the 1960s. Still what was available at that time was by no means insignificant. When a pion and a nucleon (or two nucleons) were made to collide with such enormous energies — the result was quite unbelievable.

The scientists found many new particles in the end-products of such collisions. For example, antiproton — we have mentioned this earlier, this is the antiparticle of proton. A number of other new particles were discovered besides this, about which there existed absolutely no theoretical prediction. Some of these had zero spin, some half, some one or two or more. So there were baryons as well as mesons. Altogether there appeared a complete menagerie of hadrons. By the end of the 1950s a few dozen such new particles were discovered. Now the number of hadrons has reached a few hundred.

This is actually the second complication with the hadrons. All these hundreds of hadrons — are all of them elementary particles? How terrible! We would not be quite willing to accept such a concept. It is as if we would be more comfortable if the number of elementary particles were somewhat smaller.

4.7   Gell-Mann, Zweig

In 1964, Murray Gell-Mann and George Zweig wrote two articles. The main point of both of these articles was the same. They said there is no reason to get disheartened by these hadrons. The structure of all the hadrons can be explained by a handful of objects. These objects were given the name *quarks*.

Let us give some examples. First, consider the *up* quark, written as *u* in short. This is a spin-half particle with positive charge. If the charge of a proton is denoted by Q then the charge of *u*-quark would be $\frac{2}{3}Q$, i.e., exactly two thirds of the proton's charge. Then we have the *down* or *d*-quark. It too is spin-half, but the charge is $-\frac{1}{3}Q$. So this is negatively charged. Now if we put two *u* and one *d* together, the total charge would be

$$\frac{2}{3}Q + \frac{2}{3}Q - \frac{1}{3}Q ,$$

i.e, Q or equal to the charge of the proton. So this is the proton, whose properties are similar to the combination of two *u* and one *d* quarks. Similarly, if we combine one *u* and two *d* quarks, the total charge would be,

$$\frac{2}{3}Q - \frac{1}{3}Q - \frac{1}{3}Q ,$$

i.e., zero — exactly equal to the charge of a neutron. Therefore, there is no problem to think of a neutron as a combination of one *u* and two *d* quarks.

Wait! Both the proton and the neutron are spin-half particles. On the other hand, the spin of *u* or *d* quark is again $\frac{1}{2}$. Then how could we get another spin-half particle by combining three such particles?

We need to remember that spin is actually a kind of angular momentum, in other words it is a vector. Therefore, if two of those $\frac{1}{2}$ spins point in one direction and the third in the opposite direction, the total spin would then be $\frac{1}{2}$. Evidently, amongst the two *d* and one *u* quarks in a proton — two have their spin aligned in a particular direction and the third in the opposite direction. If the spin component of the first two is $\frac{1}{2}$ and it is $-\frac{1}{2}$ for the third quark then the total spin component of the proton would be $\frac{1}{2}$ in the specified direction. Similarly, if the spin component of two quarks happen to be $-\frac{1}{2}$ and that of the third is $\frac{1}{2}$, the spin component of proton would be $-\frac{1}{2}$. Of course! The spin of a proton is $\frac{1}{2}$, and naturally it is supposed to have two components. The same situation prevails in the case of neutrons, except there we would have to consider one u and two d quarks. According to the laws of quantum mechanics, it is impossible to generate spin-half particles combining three u's or three d's together. The two particles described above are the only possibilities — the reason there exist only two kinds of nucleons.

Gell-Mann and Zweig found that to explain the structure of all the hadrons known at that time, only one other type of quark, besides u and d, was needed. This was named *strange*, or s in short. Like d the charge of s is also $-\frac{1}{3}Q$, and the spin is $\frac{1}{2}$. The difference is only in the mass. The rest mass of the s quark is about 100 to 150 MeV larger than that of the u or the d.

Inclusion of this third quark produced six more spin-half particles apart from the two nucleons. Taken together there were now eight particles, which were compared to the eight-fold way of Buddha by Gell-Mann and Ne'eman a few years earlier. All the eight particles were familiar by this time. Everyone was happy to see these explained in terms of three quarks.

However, it is quite possible for all the three quarks to have their spins aligned. In that case the total spin would be $\frac{3}{2}$. Let us see how many such particles can be there. First, without the s quark there would be,

$$(ddd),\ (ddu),\ (duu),\ (uuu).$$

All of these four particles have similar masses, because the masses of u and d quarks are very close. The charges of these particles are $-Q$, 0, $+Q$, and $+2Q$ respectively. Scientists knew about these particles too and they were collectively known as delta (Δ) particles. The charge of the first one is $-Q$, so it is called Δ^-. The second one has zero charge. To indicate that the particle is named Δ^0. Similarly, the third particles is known as Δ^+ and the fourth as Δ^{++}.

If we now consider particles with one s quark, then three particles can be found,

$$(dds), \quad (dus), \quad (uus).$$

We have said earlier that the rest mass of s is about 150 MeV larger than that of u or d. Therefore, the rest mass of these three particles would also be 150 MeV larger than the mass of the Δ particles. These three particles were also known to the scientists, and were easily identified. Similarly, if the particles contained two s quarks, their masses would be larger by another 150 MeV. Two such particles can be found —

$$(dss), \quad (uus).$$

However, if the spin is $\frac{3}{2}$, there could be particles made up of three s quarks, exactly like those made up of three u's or three d's. But no such particle was known when Yuval Ne'eman and Gell-Mann had realized that eight particles formed a multiplet of some sort, and started looking for other multiplets. So they had asserted that such a particle had to exist. Due to the presence of three s quarks, the mass of this particle would be almost 450 MeV more than that of the delta particles. And when this particle was finally discovered in 1964 it created a major uproar.

Till now we have talked about baryons with $\frac{1}{2}$ and $\frac{3}{2}$ spin. But particles with larger spins have already been observed. How could such particles be explained in terms of quarks? The spin of a quark is $\frac{1}{2}$ and a baryon contains just three quarks. The sum of the spins of three quarks could never be larger than $\frac{3}{2}$. What then?

The answer is very simple. It is possible for one quark to orbit around another quark inside a baryon. Then this orbiting quark would have angular momentum, like an electron has within an atom. When we calculate the intrinsic angular momentum of such a baryon from the outside, along with the spin of the three quarks this orbital angular momentum would also have to be included. This would then explain a long list of baryons with many different spins.

It is important to mention one important point here. From this large variety of hadrons, along with their wide range of properties, emerged a law of conservation. Something called a *baryon number* appeared to be conserved. The baryon number of each quark is $\frac{1}{3}$. And it is $-\frac{1}{3}$ for the antiparticles, i.e., for the anti-quarks. We have seen that each baryon contains three quarks. Therefore, the baryon number of each baryon is 1. Similarly, an anti-baryon would have three anti-quarks, which would be denoted by a bar overhead. An antiproton contains $\bar{u}\bar{u}\bar{d}$, and antineutron $\bar{u}\bar{d}\bar{d}$. The baryon number of these particles is therefore -1.

In radioactivity a neutron decays into a proton, an electron and an electron-antineutrino. The baryon number of a neutron is 1, same as that for a proton. And it is zero for the other two particles. Therefore, we can clearly see the conservation law being obeyed in this reaction. But what would be the situation, if instead we contemplated the decay of a proton? All the particles lighter than a proton have zero baryon number. Then if the baryon number really is conserved, the proton cannot decay. It would have to be absolutely stable. Indeed, nobody has seen a proton decay!

All mesons have zero baryon number. This means that to produce a meson we have to combine a quark with an anti-quark. For example, consider the case of the positively charged pion. This contains an u quark and a d anti-quark, i.e., $\bar{d}u$. Similarly, the negatively charged pion is $d\bar{u}$. All such combinations, including the s quark, would give rise to eight mesons — again the eight-fold way.

What are the anti-particles of mesons? A meson has a quark and an anti-quark. Its anti-particle would then have an anti-quark and a quark — that is the same combination. Take the negatively charged pion, which is made up of a u quark and a d anti-quark. The antiparticle of this would have a u anti-quark and a d quark. But we already know this particle, this is the positively charged pion. This means that we do not need to consider the 'anti-mesons' separately — the eight-fold way of mesons contains the anti-mesons too.

Now the situation has become much more comfortable. The structure of a large number of hadrons can be understood using only three quarks.

4.8   Feynman

So what is the overall situation now? What is the meaning of the statement that we can understand the structure of the hadrons using the quarks? Does this mean that the hadrons are not elementary particles but are composed of quarks? Both Gell-Mann and Zweig avoided this question.

They said that there is no need to worry about such aspects. It is as if the quarks just provide a simple strategy to understand the menagerie of hadrons. The question of whether the quarks really exist is not relevant in this context.

However, the question stopped being irrelevant within just a few years. A rather significant experiment was performed in the middle of the 1960s decade. Electrons were accelerated to very high energies and were made to be incident upon protons, in the accelerator situated in Stanford, California, USA. It is not difficult to calculate the deflection of the electron flow due to its interaction with the proton, had the proton been a solid sphere of very small radius. If, on the other hand, there exist three quarks inside a proton, then the situation would be completely different. The charge would be concentrated within the tiny regions inhabited by the quarks. These quarks, acting like stones inside a pulpy fruit, would then strongly deflect an electron; sometimes even sending it directly back. The experimental observations did hold up this view.

Certainly, those who have not forgotten the epoch-changing experiment of Rutherford can see the similarity. Rutherford had thrown alpha particles at the nucleus, and here electrons are being thrown at the protons. The stage is different, the actors too have changed but everything else have remained the same. There is no difference, the same thing is happening here too. Richard Feynman's calculations clearly demonstrated this scenario.

Therefore a proton is not an elementary particle at all, but is composed of smaller particles. Afterward, detailed experiments were carried out and a huge amount of data was collected. Analysis of this data resulted in the realisation that the properties of the constituent particles of proton are exactly like those of the quarks proposed by Gell-Mann and Zweig.

A quark, therefore is not just a strategy to solve certain problems, or a clever trick to remember the list of the hadrons — they do exist in reality. This fact was now firmly established.

4.9   Richter, Ting, Lederman

Everything was going quite well with these three quarks. But in the November of 1974, Samuel Ting's group on the eastern coast of the USA and Burton Richter's group on the west coast almost simultaneously announced a new finding. They had discovered a new meson. And it was not possible to explain this meson by the three known quarks. The only way out is to propose a fourth quark. This new quark was named *charm* or c in short. The meson observed by Ting and Richter's groups was $\bar{c}c$, a bound state of the charm quark and its antiparticle.

To be honest, the concept of a fourth quark was not undreamt of. We have discussed the eight-fold way of mesons arising from u, d and s quarks before, meaning there exist eight such mesons. Three of these are pions with three different charges, mentioned earlier. Four of the rest contain an s quark or its antiparticle. These are known as K particles. The final one can contain both s and $\bar{s}$, but we would not talk about it for the moment.

Two of the K particles are charged. One of these is positively charged and the composition is $u\bar{s}$. The other is its antiparticle, the composition being $\bar{u}s$. The other two are charge neutral, because their compositions are $d\bar{s}$ and $\bar{d}s$.

The charges of d and s are equal and opposite and therefore one of these combined with the antiparticle of either produces a neutral particle. One of these charge neutral particles is relatively longer lived and is known as K_L. The other one, known as K_S, has extremely short life.

K_L can decay in many different ways. One of the possibilities is to create a muon and an antimuon. From the three-quark model it was expected that this would not be a rare case. However, in reality it was seen that the probability of K_L decay through this process is about one part in one billion. In 1970, Sheldon Glashow, John Iliopoulos, and Luciano Maiani showed that this could be explained if a fourth quark is proposed. Then again, the masses of K_S and K_L are not exactly equal either. If there were only three quarks it would be impossible to understand why the difference is very small. In an article published in 1974, Mary Gaillard and Benjamin Lee showed that if the total number of quarks is four, that is two pairs, then they cancel out pair by pair. This can give rise to the small difference in the masses of K_S and K_L. In fact, they even predicted the mass of the new quark from this consideration. Their calculation more or less matched with the result of the experiments of Richter and Ting. Thus c quark was discovered.

In the chronicles of physics this event is referred to as the 'November Revolution'. Because now it started to look like even the quarks come in different generations. All the properties of c like charge, etc., are the same as those of u except its mass which is much larger. And we have already seen that there is no difference between d and s except for the mass. It appears that u and d constitute the first generation and c and s the second.

Does it end here or is there a third generation as well? The first mention of this appeared even before the discovery of c quark in 1974. We have just been saying that there could be many different ways of K_L decay. In 1964, Val Fitch, James Cronin, and their collaborators experimentally demonstrated that in one of these processes two pions are produced. This came as a big surprise. Because according to a conservation law, thought to be true then, this channel of decay could have been quite impossible. The experiment of Fitch and Cronin proved that the said conservation law has to be wrong. But how could one explain this unusual decay process? In 1973, Makoto Kobayashi and Toshihide Maskawa showed that it could be easily explained if the number of generations of quarks is assumed to be three instead of two.

Initially not many took this seriously, but the wind changed its direction after the discovery of third generation leptons in 1975. Then in the beginning of 1977 came the announcement from Leon Lederman and his collaborators that they have found a third generation quark. The charge of this is $-\frac{1}{3}Q$ just like d or s. This was named *bottom* or b in short. The meson seen by Lederman and his associates was $b\bar{b}$.

Now there was unanimous agreement on the fact that another quark would have to exist. Just like there are six leptons in three generations, there would be six quarks in three generations. This unborn sixth child was named *top* or t. Needless to say, this would be identical to u or c, but would be more massive.

More massive yes, but how much more nobody had any idea about. Consequently, it was not clear how this quark could be observed in the laboratory. The scientists encoun-

Particle	electric charge	baryon number
u	2/3	1/3
d	-1/3	1/3
c	2/3	1/3
s	-1/3	1/3
t	2/3	1/3
b	-1/3	1/3
$\bar{u}$	-2/3	-1/3
$\bar{d}$	1/3	-1/3
$\bar{c}$	-2/3	-1/3
$\bar{s}$	1/3	-1/3
$\bar{t}$	-2/3	-1/3
$\bar{b}$	1/3	-1/3

Table 4.2: The quarks, their charges, and baryon numbers.

tered quite a few false alarms after the discovery of b. Some of the experimental results were interpreted to be signatures of t, formal announcements were made but afterward the experiments were shown to have faulty measurements without any connection to the t quark whatsoever. In the end, all the confusion was cleared when t quark was discovered in 1994.

Finally, we have six quarks in all, and the six associated anti-quarks. All of these are spin-half particles, just like the leptons and anti-leptons. To remind ourselves about the electric charge and baryon numbers of different particles we have summarised them in Table. 4.2.

4.10   An elementary consideration

A question remains. What are the basic, fundamental constituents of the universe then? What is everything made up of?

Earlier we used to think, the atoms are the basis of everything. Then we saw that there exist electrons, protons and neutrons inside an atom. And we thought that if photons are added to this then everything can be explained.

Arriving at this chapter we find that the electrons retain their status of fundamental particles. But the lepton family also contains muon and tau and the three corresponding neutrinos. Then there are the antiparticles of each of these. And all of these are elementary particles.

But we have just seen that the muon decays into electron, mu-neutrino and electron-antineutrino. Does this not mean that a muon is made up of these three particles?

Not at all. If a muon happened to be a bound state of these three particles then energy needed to be supplied to extract these particles. The way energy is required to eject an electron from an atom. The decay of muon is not like this. Even if the muon is left alone, it would decay. This is the significance of the concept of 'decay'. This is not breaking a particle up, but the decay of it. If a particle changes into another less massive particle while obeying all the conservation laws then that is its decay, which does not raise any question about its 'elementarity'.

This is why we did not question the elementary status of a neutron, after seeing it decay in radioactivity. But then we found that the behaviour of a proton was like a composite particle when a highly energetic electron is thrown at it. Actually none of the hadrons are elementary particles, including protons and neutrons. The constituent particles of these are the six quarks and their anti-particles.

The projectiles available in Dalton's era were not energetic enough to destabilize the electrons or protons inside an atom. As a consequence Dalton considered the atoms to be indivisible. Then Rutherford performed his paradigm-shifting experiment with the alpha particles. The real nature of the atoms was unfolded. Protons were considered to be elementary particles till the mid-twentieth century. Now we know this not to be so. If we find even more energetic particles to perform Rutherford-like experiments on muon or tau and obtain similar results — then we would have to concede that the muon or tau are not elementary particles either. The same is true of electrons and quarks and neutrinos.

But that would be a question for the future, perhaps just a possibility. But if someone asks us today whether quarks and leptons are elementary particles, our answer would be yes. And this whole universe is just a manifestation of the complex variety of their actions.

Chapter 5

Destination unification

5.1 Behind the apparent

It should be obvious that a reader reading this page must be having some sort of interaction with the ink-marks that define the letters on the page. How is this interaction being established? Well, if it is evening, we assume that there would be an electric lamp glowing in the room. Light coming out of that is falling on the page, getting reflected, and being received by the reader's eyes. And the reader is reading. In short, the interaction is being established through light.

Suppose, if we ask now, how does the lamp glow? When the switch is pressed how does the lamp, sitting a few meters away from the switch, get that piece of information? We know the answer to this question. There is a wire, connecting the switch to the lamp, which carries an electric current. So in this case, the connection is established by electricity flowing through the wire.

What happens when we turn on an electric fan? There are coils of wires inside a fan. When an electric current flows through it, a magnetic field is generated around the coils. If we now insert a metallic ring within the influence of this magnetic field, the field induces a rotation on the ring. Once there is a rotating something, it is easy to fit a few blades on it so that it can send ripples in the air around it. Therefore, it is actually the magnetic field which acts as an agent in establishing the connections.

We may recall that Maxwell showed these phenomena not to be independent. Light, electricity, magnetism — these are all governed by a common set of laws. We can then summarize the statements made in the previous paragraphs by saying that two objects can interact with each other through the electromagnetic field.

We have already learned about the interior of atoms. An atom has a nucleus in its centre with some electrons going around it. How do the electrons know that there is a nucleus somewhere around? Because the nucleus contains protons and neutrons, of which the protons carry positive electric charges. These charges create an electromagnetic field around them. The electrons hover around in this field. So the electromagnetic field is acting the matchmaker again.

Sometimes when two different substances are brought close together, they react chemically. What happens in a chemical reaction? The molecules break up owing to interactions between the atomic electrons, and the atoms reorganize themselves into new molecules. Thus, here too the interaction is electromagnetic.

When a person writes, she holds a pen in her hand. How does she do that? There is something going on between the molecules that constitute the fingers of her hand which al-

low them to put a pressure on the molecules that constitute the pen. It would be difficult to describe the details because of the complexity of the process. But what is certain
is that some kind of interaction between the molecules is
responsible for her hold on the pen, and those interactions
are electromagnetic. It is the same story behind most of the
things we do — speaking, walking, sitting down, chewing
our food — anything one can think of!

But if the pen slips out of her hand and falls to the floor,
that is not due to any electromagnetic interaction. Here the
gravitation of the Earth is responsible for the phenomenon.
Just as a charged particle or a magnet sets up an electromagnetic field around it, a massive particle sets up a gravitational field around it. Because of the gravitational field the
pen in her hand would know the presence of the Earth near
it. So, as soon as it slips out of her hand, it goes down and
reaches the floor. So here the interaction is gravitational.

The Sun produces a gravitational field around it. The
planets experience this and orbit around the Sun as a result. The stars and galaxies in the sky are moving in various
ways, all because of gravitation. The moon is encircling the
Earth. How does the moon know about the Earth? Again,
because of the Earth's gravitational field.

Arguably we are talking about the most fundamental
question of physics. How does any object know about other
objects? How does any object relate to any other? How does
one object influence another? How does an object behave
under such influences?

If nothing like this happened, if there were no interaction between the objects of our Universe, there would be
nothing to discuss in physics. In fact, there would be no one
to discuss physics with, or anything else, either. Because

our body is made out of conglomeration of molecules, and the organization and function of these molecules depend crucially on the interaction between them. Without these interactions, nothing could form — no sun, no planet, no plant, no insect, no nothing.

Almost every phenomenon that we can see with our naked eyes or feel with our other senses can ultimately be explained with only two kinds of interactions: electromagnetic and gravitational.

5.2   Beyond the apparent

If we try to understand the phenomena that take place at scales which are too small for us to see, we realize that only those two interactions, mentioned before, are insufficient. Take, for example, the nucleus of an atom. There are protons and neutrons in a nucleus. The gravitational attraction between them is insignificant, the magnitude of the repulsion between the protons is much stronger than that. Still the nucleus does not get torn apart. The question is why.

We have already explained that the reason for this is the strong force. Any proton or neutron creates a field of this force around it. This is called the field of the *strong force*. Through its effects, i.e., through strong interactions, the net force between protons and neutrons in a nucleus is attractive. And this is why we can obtain nuclei which are long-lived.

Does it mean that all nuclei live happily ever after, being bound by the strong force? No, that is not the case either. We know about the phenomenon of radioactivity. It is basically a class of phenomena in which nuclei spontaneously break up, emitting certain particles in the process. A substantial amount of a radioactive material would then emit a large number of such particles from its nuclei. The flow of these particles are collectively known as a *radioactive ray*, which has wide-ranging applications today from medicine to many other fields.

There are different types of radioactivity. In one type, a neutron in a nucleus decays to give a proton, an electron and an electron-antineutrino. How does that happen? Strong interaction cannot be responsible for this, because strong interaction has nothing to do with the electron. Could it be possible then that the electromagnetic repulsion that we talked about earlier is somehow dominant in these nuclei? But that cannot be the case either, because electromagnetic interaction cannot change a neutron to a proton.

Not strong, neither is it electromagnetic. Gravitation is negligibly small at these scales. So there must be a fourth kind of force. That force must be weaker than the strong or the electromagnetic force, which is why all nuclei do not disintegrate. And for this reason, we can call this force the *weak force*.

It should not be concluded, from what has been said above, that strong force manifests itself only through neutrons and protons, or weak force through radioactive nuclei. There are a host of other phenomena in which these forces play crucial roles. How do electron, mu-neutrino and electron-antineutrino appear as a result of a muon decay? There is no question of the strong force playing any role here. Because none of the particles involved is a hadron and consequently has nothing to do with the strong force. Neither is the electromagnetic field suited for the job, as mentioned earlier. So it is the weak force again, responsible for muon decay. Similarly, the tau decay and many other such processes are controlled by weak interaction. Many of them include other kinds of particles than the ones we have mentioned so far.

In summary, there are four kinds of forces, or four kinds of interactions. The strongest one is known, not surprisingly, as the "strong interaction". Electromagnetic interaction is next in the order of strength, followed by weak interaction. And gravitational interaction is so feeble that we will ignore it for the rest of this book entirely.

5.3 Give and take

A discussion on forces inevitably leads us to the fields. Maxwell was the first person to talk about electromagnetic fields in the nineteenth century. The force of gravitation was known since Newton's time, though the equations governing the gravitational field came into existence only at the beginning of the twentieth century — through the efforts of Albert Einstein. All this happened before quantum mechanics was developed. However if the details of an atomic nucleus were known in Maxwell's time, had radioactivity been discovered — perhaps we would have learned about the strong or the weak fields even before the quantum theory came to be established. Perhaps such a field theory would have explained how waves are generated in a force field, how energy is transported through such waves and connection between different objects is established through them.

However, we have become more knowledgeable now. Such old fashioned theories no longer do. What is the need

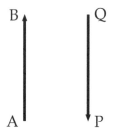

Figure 5.1: Two current-carrying wires. The arrows indicate the directions of currents.

for saying that energy is transferred through waves? We already know that many of the properties of an electromagnetic field can be explained by a type of particles. We do not even need to be reminded that those particles are the photons.

The discussion on photoelectricity showed us that when light is incident upon metal, electrons are ejected out of it. To explain this we said that the particles of light, the photons, are colliding with the electrons — exactly like a cricket bat hitting the ball — and as a result of that collision the electrons are being thrown out of the metal. An interaction of the electron with the electromagnetic field is explained using the photons in this way.

But this is not enough. Consider two wires, carrying electrical currents in the directions shown by arrows in Fig. 5.1. A repulsive force would be felt between the wires. How does that happen? The two wires are not touching each other and no information can flow from one wire to the other. How does one wire then know about the other wire?

A classical physicist would follow Maxwell's theory to answer this question. He would say that when a current is flowing through the wire AB, it creates an electromagnetic field around the wire. The waves associated with this field carry the information from one wire to the other. The wire PQ resides within this electromagnetic field, and thus comes to know about the wire AB. The wire AB learns about the wire PQ in exactly the same manner: through the electromagnetic field.

That is fine, but we would like to understand this phenomenon from a quantum mechanical point of view, without involving the concept of waves in the electromagnetic field. We would like to have a description in terms of the photons. Now we have a different story to tell. Let us start by thinking about the electric current flowing through a wire. There are electrons in the material from which the wire has been constructed. These electrons are flowing. Now, in the course of this flow, an electron in the wire AB could emit a photon. Perhaps this photon travels in the direction of the other wire and collides with it. This photon could then carry a bit of information from the wire AB to the wire PQ.

It does not mean that a specific electron in the wire AB emits only one photon. There would be many electrons flowing through the wire, and each of them would be emitting a photon once in a while. Not all of them are emitted in the direction of the wire PQ. They are emitted in all directions. So, the space around the wire would be teeming with photons at any given instant of time. The collection of these photons is what a classical physicist would call an electromagnetic field.

Among these photons, some travel in the direction of the wire PQ. The wire PQ would absorb them. It cannot possibly absorb all photons that come its way, but would

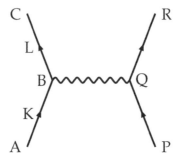

Figure 5.2: The simplest Feynman diagram depicting interaction between two electrons.

be able to catch some of them. Similarly, the wire PQ would be emitting a large number of photons, and the other wire would be absorbing some of them. The exchange of these photons would establish the interaction between the two wires.

It is not really necessary to think about something as complicated as two wires, with zillions of electrons flowing through them. We can think of just two electrons. Each of these two will somehow feel the effect of the other. And the reason would be the same. One electron would emit some photons which the other one would absorb, and vice-versa.

Of course, a picture is worth more than a thousand words. Richard Feynman showed how to summarize all such words in simple diagrams. In Fig. 5.2, an electron is traveling along the path AKBLC. Along the way, it emits a photon when it reaches the point B. The other electron, starting from the point P, catches this photon at the point Q.

Such pictures are called *Feynman diagrams*, and they should not be thought of as photographic depictions of the real events. In Fig. 5.2, we do not say that the path of the electron has a sharp bend at the point B. The photon in the

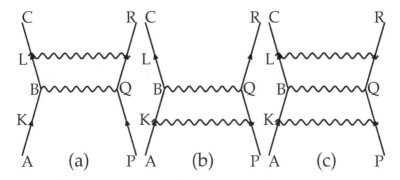

Figure 5.3: Examples of more complicated Feynman diagrams depicting interaction between two electrons.

middle is obviously not traveling along a wavy line. All lines in the figure are symbolic: the wavy line represents the photon, the straight lines represent electrons. The important message that is given in the figure is this: only one photon has been exchanged between two electrons.

We might ask, why did the electron emit a photon precisely at the point B? Could it not have emitted a photon when it was at K? Or at L? What is so special about B? Similarly, why did the other electron catch the photon at the point Q? What is special about this point either?

Nothing. Nothing special about these points at all. In fact, the first electron could have emitted a photon at any point. The probability is non-zero throughout the path of the electron. The photon could have been emitted at K or L or any other point. The point B is just incidental. Similarly, it is a matter of chance, or probability, that the other electron caught the photon at the point Q.

Since the probability exists, could it not have happened that one photon was emitted at B, and another one at L, and the other electron caught both of them? Well, yes, it could have, and Feynman would have represented it by the diagram shown in Fig. 5.3a. The second photon could have been exchanged between two different points as well, as shown in Fig. 5.3b. Note the word "exchanged". We are no longer saying which electron emitted a photon and which one absorbed it. In all of these pictures, we can also think of the photon being emitted by the second electron and caught by the first electron. It is irrelevant which particle emits the photon and which absorbs it. The important fact is the exchange of photons. Many such exchanges can also take place, as shown in Fig. 5.3c. While calculating the force between two electrons, all such diagrams will contribute. We will have to add all of them up in order to obtain the total interaction between two electrons.

Planck and Einstein, in the first few years of the twentieth century, showed how photons can explain various phenomena where light directly interacts with matter. Take, for example, the case of the photoelectric effect, which happens because of scattering between light and electrons. So light, or the electromagnetic field, is the actor in this play. And now we go one step further and see that the electromagnetic field can do more. It can function like the director of a play, who remains behind the wings, but decides how different actors should interact with one another. In other words, now we can describe all the properties of an electromagnetic field in terms of photons. This language, or manner, of description is called *quantum field theory*. In this language, an electromagnetic field is a collection of many photons. Or, turning it around, we can say that a photon is the quantum, or the particle, of the electromagnetic field.

What would have happened if, instead of electrons, we had some other particles? We said that at any instant, there is a probability of the electron emitting a photon. The same would be true for any other charged particle. Suppose we consider two d quarks, each having a charge equal to one third that of the electron. In the language of quantum field theory, it means that at any instant, the probability of a d quark emitting a photon would be one third the probability of an electron emitting a photon. The probability of absorbing a photon also suffers from the same factor. Thus, if we think that the solid lines in Fig. 5.2 correspond to d quarks rather than electrons, we would have one third of a chance that the quark will emit a photon at the point B, and a further one third of a chance that this photon will be absorbed at the point Q. So, the process in its entirety would have a probability of $\frac{1}{3} \times \frac{1}{3}$, or $\frac{1}{9}$, compared to the same process with electrons. The repulsion between two d quarks would therefore be one ninth that of the repulsion between two electrons. This is what Charles Coulomb had discovered in the 18th century: the magnitude of the force between two charged particles is proportional to the product of the charges of the particles. More complicated diagrams, like those appearing in Fig. 5.3, would yield different ratios, but then the contribution of these diagrams are so small to begin with that they hardly matter.

We seem to be getting back all the results of classical electromagnetic theory through this new language, sometimes with a small correction that goes unnoticed in the classical version. Let us now ask the question that would prove extremely important in what follows. How do we obtain the law of conservation of charges in the language of quantum field theory?

The answer is very simple. Let us go back to Fig. 5.2 one more time. What is happening at the point B? An electron emits a photon. The charge of the electron does not change in the process, of course. Thus no change of charge will take place in the process if the photon is considered to have zero charge. There is no problem with charge conservation if an electron emits a photon. There is no problem if an electron absorbs a photon. There is no problem if the charged particle is not electron but something else. There is no problem if the particle emits a hundred photons and absorbs seventeen.

So we have charge conservation. What happens to energy or momentum conservation?

We need to mention something here. We have been discussing the concepts of the quantum mechanics, the quantum field theory and so on. Heisenberg's uncertainty principle has been at the root of all. It says that the position and the momentum of a particle can not be simultaneously measured to any arbitrary precision. This is not due to any limitation in our measuring instruments but because the nature defines it this way.

How do we know that the momentum is conserved? We need to measure the momentum before and after an event. If the two momenta measured are exactly equal only then we can say that it is conserved. But if there exists any uncertainty in our measurement then it is impossible to say anything about conservation. For example, let us say that two particles are colliding 'inside this room'. The moment we say 'inside this room', it becomes apparent that we know something about the position of the particles. However precise the measuring instrument be, we cannot therefore make precise measurements of the momenta of the particles. Perhaps the first measurement gives a result between 7 and 9 in a particular unit and the next one between 6 and

8 units. If the momentum is actually between 7 and 8 units then it does fall within the range of both measurements. So it cannot be said that the momentum is not conserved. But neither can we say that it is being conserved with any conviction. It is quite possible that the change in momentum is within the uncertainty of measurement. Then it is neither possible nor correct to say anything definite about the conservation of momentum.

The energy and the time are also intimately related to each other, exactly like the position and the momentum. Therefore, we cannot talk about energy conservation either, in quantum mechanics. Energy may not be conserved, but only within a certain time interval. If the initial energy of a system is E, then its final energy could be $E+e$ even if there is no exchange of energy with the outside. But this state of affairs can persist only for a time interval t such that the product of e and t would be about $h/4\pi$ — just like in the case of position and momentum.

We need to remember this about the photons in the Feynman diagrams too. A simple calculation would show that if the momentum or energy were conserved precisely then an electron would not have been able to emit a photon. The photon has been emitted violating the energy conservation. Therefore it has a limited lifetime, defined by the uncertainty principle as discussed above. Such photons are called *virtual* photons. All the photons in Fig. 5.2 and Fig. 5.3 are such virtual photons.

That is the difference of these processes with the photoelectric effect. The photons, incident upon a metal in photoelectricity, have not been produced through a violation of energy conservation. Those photons are therefore real, not virtual. Maxwell's field theory is not of much use where the photons are real. That is what happens in the case of photo-electricity. The photons have to be thought of as particles there.

However, ideas of classical physics can be used where the photons are virtual. That is how the mutual repulsion of two electrons can be explained by Maxwell's field theory. The conglomeration of the virtual photons around an electron is the electromagnetic field. We have discussed the field-particle duality in Chapter 2. In terms of the quantum field theory everything can be described as particles — some real, some virtual.

5.4   The symmetries

Near the end of Chapter 1 we said that every conservation law has an inherent symmetry associated with it. The renowned mathematician Emmy Noether showed this in the nineteenth century. For example, the law of conservation of momentum can be derived from translational symmetry of space, i.e., the hypothesis that no point in space is special, and we can set up the origin of a co-ordinate system anywhere we please, with identical consequences. Energy conservation can be derived from the homogeneity of time, and rotational symmetry of space gives the conservation of angular momentum.

We also talked about the charge conservation in Chapter 1. But we did not really address the question of the symmetry associated with it. To answer that we need to look at something else first.

Schrödinger used a symbol ψ to denote the matter waves. We have said that if the magnitude of a physical quantity varies with space and time, it can be said that there exists a wave associated with that quantity. For electromagnetic waves, the electric field, or equivalently the magnetic field, varies in space and time. The matter wave is nothing but the variation in ψ. But there is a difference. The height of water level or the electric field are represented by ordinary numbers. The matter wave amplitude ψ, on the other hand, is represented by somewhat more complicated things called *complex numbers*. For the purpose of our discussion, we can think of a complex number as an arrow on a piece of paper, i.e., an arrow in two dimensions. These are not physical dimensions like length, breadth or height; these are some hypothetical directions. The value of ψ at any point at any instant can be represented by an arrow at that point at that instant. The length of the arrow would represent the magnitude of ψ, and the direction of the arrow would represent the direction of ψ in that hypothetical space embodying complex numbers. The length is called the *modulus* of the complex number, and the direction its *phase*.

What does ψ mean? Earlier, in Chapter 3 we said that if ψ represents a particle at a point, then the square of the length of the associated arrow is the probability of finding the particle at that point. Good, but not enough. We also need to find out the physical implications of the direction of the arrow.

For that, let us suppose that two streams of electrons are coming from two points A and B, falling on a screen, as seen in Fig. 5.4. If only the beam from A came to a point, the matter wave amplitude would have been ψ_A. If it came only from B, the amplitude would have been ψ_B. When

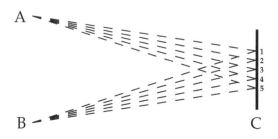

Figure 5.4: Two streams of electrons falling on a screen.

both of them come together, the amplitude would then be the sum of the two, i.e., $\psi = \psi_A + \psi_B$.

But we have to remember that these objects that we called ψ or ψ_A or ψ_B are not ordinary numbers. They are complex numbers or arrows. So, if at a point we have the arrows corresponding to ψ_A and ψ_B which are equal in length but opposite in direction, the effects of the two would cancel at that point and the sum, i.e., ψ, would be zero at that point. It would mean that at that point, there is no way we can find an electron, no matter how hard we look for it. If we look at Table. 5.1, we find that such are the cases at the points marked 2 and 4. On the other hand, if the arrows corresponding to ψ_A and ψ_B point in the same direction somewhere, they would reinforce each other, and the probability of finding an electron would be higher there. This is what happens at the points 1, 3 and 5 of Table. 5.1. If the directions of the two arrows are neither the same nor the opposite, the sum will be somewhere in between. In summary, if we have two streams of electrons falling in a region, there will be points where we will see a lot of electrons, and other points where we will see less, and some points where we will see none.

	ψ_A	ψ_B	$\psi_A + \psi_B$
1	→	→	⇒
2	↓	↑	.
3	←	←	⇐
4	↑	↓	.
5	→	→	⇒

Table 5.1: Arrows represent the phases of electrons coming to different points shown in Fig. 5.4. The source points for the electrons have been shown as a subscript on ψ. A dot on the rightmost column represents a cancellation between the two contributions, whereas a double-lined arrow represents reinforcement.

Such a phenomenon is called *interference*, and is expected of any wave. We mentioned this in Chapter 2. It was known to happen to light waves for a long time. Davisson, Germer and Thomson deduced the wave nature of electrons from observations of such interference of electron waves. For matter waves, the evidence was obtained in the first half of the twentieth century.

Let us think about the whole thing. Suppose we perform such an experiment. We find the place where the probability of finding the electrons is zero. We know that at these places, the arrows for the two streams are in the opposite directions. In more formal language, we know that the phases are equal and opposite to each other. True, but do we know the individual phases? The answer is 'no'. We know that in the hypothetical space where phases are like directions, if ψ_A points along the direction of 10 on the face of a clock, ψ_B points towards 4. If ψ_A points towards 6, ψ_B towards 12, and so on. We do not know more than that. Similarly, at places where the probability of seeing the electrons is highest, we know that the arrows corresponding to ψ_A and ψ_B point in the same direction, i.e., the phases are the same. But we do not know what that phase is.

Imagine that we are all asleep at a time when some genies appear and rotate the directions of these arrows everywhere in the universe. Would it make any difference? If we have two arrows and we rotate both of them by the same amount, the angle between them would not change. If they used to be back-to-back, they would remain so. If they were in the same direction, they would continue to be in the same direction after the genies' work. That would mean that the probabilities of finding the electron at different points would remain the same even after this change. We would not be able to suspect that the genies have made some change.

Now we get a hint of a symmetry. If a change of a particular quantity does not produce any change in another quantity, it is called a *symmetry*. We have talked about the translational symmetry of space and time earlier. Here we encounter the symmetry of the phases. The relative phase between the two streams of electrons is important, and we can see its consequences. But if all the phases are changed by the same amount, that would not have any effect on the physical universe.

But wait, there is a caveat! The phases must be changed by the same amount everywhere. If instead we change the phase by different amounts at different places, the result will be appreciable. For example, let us go back to the two streams of electrons shown in Fig. 5.4. And suppose we change the phase of the stream coming from A by a half turn, doing nothing to the stream at B. In Table. 5.2, we have shown what will happen now when the two streams meet.

	ψ_A	ψ_B	$\psi_A + \psi_B$
1	←	→	·
2	↑	↑	⇑
3	→	←	·
4	↓	↓	⇓
5	←	→	·

Table 5.2: Same as in Table. 5.1, except that an extra half rotation has been introduced for electrons coming from the point A.

Previously, the arrows at the points 2 and 4 were back-to-back. Now, they would be pointing in the same direction. In practical terms, it means that now we would get the maximum number of electrons at a place where we failed to find any electron earlier.

Obviously, the genies need not do something as dramatic as giving a half turn to the phase at one point in order for it to be experienced. Even if the change of phase deviates by an arbitrarily small amount from exact equality anywhere, the difference in the interference pattern would be discernible. Turning things around, we can say that as soon as we see a change of interference pattern, we would know that the phase has changed somewhere: the frivolity of the genies would be exposed.

But these genies do not want to be exposed, so they have arranged a deep conspiracy. The point is that, if an electron had emitted or absorbed a photon at the point A, it also would have changed the phase of the electron. That would have caused changes in the interference pattern if the electron had met another electron subsequently.

It means that, if we see a difference in the interference pattern, we cannot immediately conclude that the arrow of ψ has been rotated somewhere. We should be aware that the difference can just as well be caused by the electron emitting or absorbing a photon on the way.

So the bottom line is the following. Earlier, we said that if the genies changed the phases of everything by the same amount everywhere, we could not have known that. Now, we find that even if the phases are not changed by the same amount everywhere, there is no way for us to know that.

This kind of calculated confusion is called *gauge symmetry*. The name is bad, and makes no sense, because the word 'gauge', in English, means a measuring instrument, as in 'rain gauge'. But if a meaningless concoction somehow receives universal acceptance, we cannot but go along with it and think, "what's in a name!" Certainly a name such as *phase symmetry* would have been much more appropriate, but who is listening?

Anyway, let's go back to the question that appeared near the beginning of this section. It was a question about the symmetry behind the law of conservation of charge. Well, the answer should be obvious now. The symmetry behind this conservation law is the gauge symmetry that we described in this section.

5.5   Old wine in a new bottle

Let us repeat what we just said about gauge symmetry. The genies want to change the phases, or the arrows. If they could do that by the same amount everywhere at the same instant, we could not have possibly noticed their work. But that's easier said than done! Just imagine: they will have to change the phase in Kolkata, in Hyderabad, in Paris, in Abidjan and in Hanoi, all by the same amount, at the same time. They will have to do the same behind the clouds, near the sun, away in the galaxies. Oh, that's too much even for a genie! In a more serious tone, we can say that the tenets of the special theory of relativity does not even allow such an operation.

Of course there is nothing against changing the phase in a small region. The genies can do that. But they are afraid that we will get to know what they are doing. So they have devised a particle called 'photon'. Because there are photons, we cannot really tell whether the phases are being changed.

Of course, as we learned earlier, photons are just carriers of electromagnetic interactions. Thus we can say that the electromagnetic interactions are results of gauge symmetry. It is a façade to hide the undercurrents of phase rotations.

We are saying the same thing that we said in the last section, but from a different point of view. This is how Chen-Ning Yang and Robert Mills described the interactions in 1954. With this new interpretation they could generalize the idea to other kinds of symmetries and hinted that one should try to explain other interactions with such generalizations as well.

There are three other kinds of interactions, as we described earlier. Barely about a decade and a half after the Yang-Mills prescription, it was seen that weak interactions can be understood through gauge symmetries. And then, in 1974, it was realized that strong interactions can also be explained the same way. The gauge theory of strong, weak and electromagnetic interactions constitute what is known as the *standard model* of interactions.

Notice that we have left out gravitation. We will comment on it later. Right now, our aim should be to try to understand the standard model, i.e., to understand how gauge theories explain strong and weak interactions. Historically the mystery of weak interactions was unraveled earlier. But we will take an anachronistic approach and describe strong interactions first, for reasons to become obvious as we proceed.

5.6　The "colors"

The particles interacting via strong force are called hadrons, like proton, neutron, pion or delta. In the last chapter we have seen that these hadrons are made up of six types of quarks and their corresponding antiparticles.

For example, let us consider the delta particles. There could be four kinds of delta particles. One of them is Δ^- which has a total charge of $-Q$, since it contains three d quarks. So its charge is exactly opposite to that of the proton. On the other hand a Δ^{++} contains three u quarks and has a total charge of $2Q$. The other two delta particles contain both u and d quarks.

To understand the behaviour of electrons inside an atom we used the exclusion principle of Pauli, which prevents two electrons to be in the same quantum state. Protons within a nucleus also follow similar rule. This implies that all spin-half particles would obey the exclusion principle.

But the quarks are also spin-half particles. What is happening with them? Δ^- has three d quarks and they should be in different quantum states. And yet, the spin of Δ^- is $\frac{3}{2}$ which means that all the quarks have spins in the same direction. And the orbital angular momentum is zero for all of these. This means that they are in the same quantum state. How is the exclusion principle satisfied then?

It is possible that we have not interpreted the quantum states of these quarks correctly. The quantum state of an electron inside an atom is defined by its energy, orbital angular momentum, spin angular momentum and the components of these angular momenta along a chosen direction. It is not sufficient to know just these quantities for a quark. It is also necessary to know the *color* of a quark.

Quarks can come in three colors: red, blue and green. When we say that, it must be noted that we do not mean that the quarks share the same visual characteristic as the setting sun or the autumn sky or the leaves of a tree. The 'color' represents a new property of matter here, and has no connection with the sense in which we use the word in everyday language.

Whatever may this property be, it saves the exclusion principle for us. As we said, a Δ^- contains three d quarks. The three would be in the same quantum state if color is not accounted for. So we now have one of the quarks to be red, one blue, and the other green. This would ensure that each quark is in a different quantum state. Same can be said about Δ^{++}. It contains three u quarks, but each with a different color.

Now consider there are genies who are trying to confuse us about this novel property called color. They are changing the colors of everything that is colored. Are we going to know about it? If they change all colors consistently at the same time, we would not know. In a Δ^- particle, if the ge-

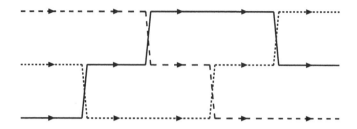

Figure 5.5: Binding of quarks in a hadron by exchange of gluons. The three horizontal lines represent three quarks in a baryon, with different dash patterns representing different colors. The cross-shaped connectors between the lines represent gluons.

nies changes the red quark to blue, the blue quark to green, and the green to red, there would still be one red, one blue and one green quark in the Δ^-, and we would not face any problem with the exclusion principle or anything else.

But we discussed earlier that these genies are not that efficient. Rather, they cannot be. They cannot change the colors of all quarks everywhere in the same way. Suppose their activities have been limited to the place where there used to be a red quark in a Δ^-, and they have changed it to blue. Since there was a blue quark to start with, this change would cause a problem with the exclusion principle. And if that happens, we would know what the genies have been trying to do surreptitiously. We see such a situation in Fig. 5.5, where different colors have been represented by different style of dashing the line representing quarks.

But the genies would not allow us that pleasure. So they have invented some new kinds of particles called *gluons*, which play the same role that the photons play in electromagnetic interactions. A quark emitting or absorbing a

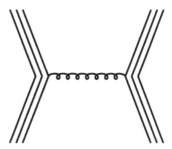

Figure 5.6: Gluon exchange mediates strong interactions. The three solid lines on each side are supposed to represent three quarks in a particle like the proton or the Δ's. The exchanged line represents a gluon.

gluon can change color. Thus, a red quark can change into blue by emitting a gluon, and another quark might change from blue to red by absorbing the same gluon. Exchange of gluons maintain the color, and this is the way that strong interaction is mediated. A schematic figure is given in Fig. 5.6.

This, by the way, is a gauge symmetry, though with a difference. In the case of electromagnetism, we commented that the emission or absorption of a photon does not change the charge of a particle. In the present case, we said that the emission of a gluon, for example, can change the color of a quark. The emitted gluon carries this information and dumps it on another quark, which then changes color accordingly. Thus, there can be different kinds of gluons. For example, one kind can be called $r\bar{b}$, meaning that if such a gluon is emitted, it can turn a red quark into a blue quark. When it is absorbed, it does the opposite thing of course, i.e., it can turn a blue quark into red. Similarly, there would be $b\bar{r}$ gluons, $r\bar{g}$ gluons, and so on. There will be eight kinds in all.

In the mid-1970s, it was hypothesized that the exchange of these gluons is the mechanism by which strong interaction operates. The gauge theory describing strong interactions in this way came to be known as *quantum chromodynamics*: 'khroma' means color in Greek.

5.7   The pictorial description

What can we say about weak interactions? Can it be described by a gauge symmetry as well? We have explained electromagnetic interactions by exchange of photons. Similarly, strong interactions are mediated by exchange of gluons. Which particles play the corresponding role for weak interactions?

Photons and gluons have spin, and the value is 1 for both in the unit in which we are specifying all spins. It is assumed that the mediators of weak interactions should also have spin 1. But, unlike photons, these particles cannot be uncharged. The charge of the hypothetical particle, in fact, is equal to the charge of the proton. The particle does not even have a full proper name: only the letter W (for 'weak', presumably) is used to denote it. Since its charge is positive like that of the proton, W^+ is a more explicit name. It is known that, for every particle there must be an antiparticle with opposite charge. Thus, corresponding to the W^+, there is also a negatively charged W^-.

Let us see how these particles help us understand the decay of a muon. The process has been shown in Fig. 5.7. The muon has emitted a W^- particle. The charge of the muon, in units of proton charge, is -1, same as the charge of the W^-. Therefore, after emitting the W^-, the muon cannot

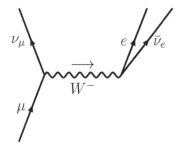

Figure 5.7: Decay of the muon. The results of the decay are the electron, the mu-neutrino (ν_μ) and the e-antineutrino ($\bar{\nu}_e$).

remain a muon: it must turn into a neutral particle. This is the mu-neutrino or ν_μ. And the W^-, after a while, has turned into an electron and an electron-antineutrino. Thus, one obtains three particles in the decay of a muon.

What happens in the case of β-radioactivity? Now we need to look at Fig. 5.8. Basically, β-radioactivity means the decay of a neutron into a proton, an electron and an electron-antineutrino. The neutron contains three quarks — one u-quark and two d quarks. If one of these d quarks gets metamorphosed into a u-quark, we will obtain a particle with two u quarks and a d-quark, which would be the proton. And how can this metamorphosis take place? Well, through the emission of a W^- particle. This W^-, as in the example of the muon decay, creates an electron and an antineutrino, and that is how we obtain β-radioactivity.

These pictures for weak interactions look very much like the corresponding pictures for electromagnetic or strong interactions. Instead of the photon or the gluon, we have the W, which is the only important difference. But then why are weak interactions weak?

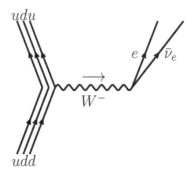

Figure 5.8: Decay of the neutron. The results of the decay are the proton, the electron, and the e-antineutrino.

The answer that was forwarded was this: the W particles are very heavy. This is in sharp contrast with what we had for strong or electromagnetic interactions. Gluons are all massless, so is the photon: their energy is all kinetic. Because they are massless, it is easy to emit and absorb them. For the W, since the mass is large, the same processes are very much inhibited.

Let us be a bit more explicit. In Fig. 5.8, we see a proton and a W particle being produced at a point where a neutron is being annihilated. Suppose the initial neutron is at rest. Its kinetic energy is therefore zero. It would still have its rest energy which is about 940 MeV. If energy has to be conserved, the total energy of the proton and the W together should also be 940 MeV then. But this is clearly not possible, since, as we know now, the rest energy of W is roughly equal to 81000 MeV. Even if we forget about the energy of the proton and the possible kinetic energy of W, there already is a big mismatch.

In classical physics, this would have spelled the impossibility of this event. Not so in quantum theory. Note that

W is not produced as a physical particle in the process: it only appears as an intermediate state. So this W is a virtual particle and the conservation of energy fails here. But the mismatch is extremely large and the process very rare. As we noted earlier, it is very difficult to emit or absorb a W particle. This is why the weak interaction is weak.

An analogy might help. Suppose the residents of a locality decided that if anyone makes a surprise visit to someone else's house and finds no one at home, the visitor must leave a card carrying his or her name, so that the residents of that home get to know who visited them while they were away.

In another part of the world and in another civilization, the use of paper is unknown: they could write only on stone tablets. They had the same idea of leaving a 'visiting card', but in their case, they had to carry stone tablets with them whenever they wanted to pay a visit to anyone else.

It will be trivial to guess which community of people has more interaction among its members. Photons and gluons are like paper cards, and W particles are like stone tablets. No wonder that weak interactions are so feeble!

But we discussed that photons are required by gauge symmetry of electric charge, gluons are required by gauge symmetry involving color. Can we not mandate the W by some similar gauge symmetry?

There is a problem though. If we set up a gauge symmetry in the manner that Yang and Mills showed us, and then introduce some particles as guardian angels of that symmetry, these new particles ought to be massless like the photon or the gluons. But we just said that the W particles are very massive. Hmm, we have a case at hand!

5.8   Subtle is the Lord

Indeed, it is true: any gauge symmetry dictates that the mediating particles associated with it should be massless. The question is: do we always see things that 'should' happen? There are many things which ought to vanish because of some symmetry, and yet they don't. Take, for example, the case of the magnets. If we take a piece of iron and rub it with a magnet along a specified direction, the piece of iron turns into a magnet. Why does that happen? Let us start from the question of why it does not happen with an ordinary piece of iron. Each atom inside the piece of iron is a minuscule magnet. But normally, the axes of such magnets are oriented randomly, as shown in the left panel of Fig. 5.9. If a second piece of iron is brought close to this piece of iron, each tiny magnet will try to attract this piece in the direction of its axis. Since the axes of the magnets are randomly oriented, so will be the forces, and their effects will cancel out. As a result, we will see that the piece of iron will not behave as a magnet. Once we rub the piece with another magnet, the internal magnets all get aligned, as shown on the right panel of Fig. 5.9. In this case, all atomic magnets will pull a nearby piece of iron in the same direction, and we will conclude that we have a magnet at hand.

Now imagine a Lilliputian scientist sitting inside this magnet. What will he or she observe? All the atomic magnets around the scientist are pointing in the same direction. So the scientist might think that this direction is special compared to others.

We who know the entire story, would not agree. We think that Mother Nature does not prefer any direction over the other. The atomic magnets inside the piece of iron are

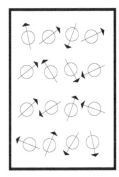 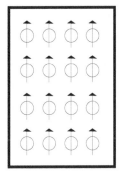

Figure 5.9: Orientation of little magnets in a piece of iron.

oriented in a particular direction because we rubbed the piece along that direction. We have picked one direction over others by rubbing it in that particular direction. Had we rubbed along some other direction, magnetism would have appeared in that direction.

If we had not chosen any particular direction and rubbed the piece of iron along it, the atomic magnets would have remained randomly oriented, and the total magnetization, summed over all those randomly directed objects, would have been zero. Turning things around, we can say that if the magnetization were zero, the rotational symmetry in the laws of nature would have been apparent even to the Lilliputian scientist sitting inside the piece of iron.

But symmetries are not always so conspicuous. Their faces are sometimes hidden under veils. This is exactly what happens when the piece of iron is magnetized. We could rub the piece along any direction. The piece would have developed magnetization along that direction, no matter what the direction was. There is symmetry in this respect, and any direction is equivalent. However, the fact remains that we rub along some chosen direction, and magnetization develops along that direction. As a result, symmetry has become hidden. It has made things difficult for the Lilliputian scientist: he or she cannot see that Nature has no preference as far as directions are concerned.

A similar thing is happening when we, the not-so-Lilliputians, are trying to think about the W particles which mediate weak interactions. There is a gauge symmetry which says that the mass of the W should be zero, just as rotational symmetry says that magnetization should be zero. But we are sitting in a universe where that symmetry is not plainly apparent. With a hidden symmetry, the W particle can have mass, just like a piece of iron can have magnetization.

The physical ideas behind these things were developed through the works of many scientists in the 1950s and the 1960s. In 1967, Steven Weinberg created a gauge theory based on such ideas. A few months later, Abdus Salam also independently hit upon the same idea.

And this initiated a kind of a revolution in physics. We said earlier that the gauge theory of strong interactions was discovered a few years later. Thus, with the announcement of the gauge theory of weak interactions, it was realized that gauge theories could also be useful in describing interactions other than electromagnetic. Of course, the confidence in this theory did not come overnight. In fact, most people doubted the mathematical viability of the theory of hidden symmetries until, in 1971, Gerard 't Hooft removed such doubts in a brilliant set of papers.

The theory of Weinberg and Salam is based on the existence of some particles which were unknown at the time the theory was proposed. First of all, they said that it was not W alone, but there exists another particle that acts as the mediator of the weak interactions. This came to be known as the Z. The Z's are somewhat heavier than the W's. Like photons, the Z particles do not carry any electric charge. But, unlike photons, it can be emitted and absorbed even by uncharged particles like neutrinos. Thus, neutrino interactions provide the best testing ground for this Z particle. And indeed that was what happened: some interactions involving neutrinos were observed in 1973 which could not have been mediated by a charged particle like the W: they could only be the result of Z mediation. A decade later, a huge group of scientists, working under the leadership of Carlo Rubbia, detected the W and Z particles directly. In other words, they observed processes in which the W and the Z participated not as mediators which are virtual particles, but as real particles appearing in a physical process.

The other crucial ingredient of the Weinberg-Salam theory is a spin-0 particle that should be left over in the mechanism that provides masses for the W and the Z particles. The mechanism, in some pristine form, was proposed by several scientists, as we mentioned a little while ago. One of these scientists was Peter Higgs, and the spinless particle of the Weinberg-Salam theory has come to be known as the *Higgs boson* after him. Admittedly, the name does injustice to scientists like François Englert and Robert Brout, who came up with the same idea at the same time, and to some others who had other closely related ideas. However, notwithstanding the problem of giving proper credits, the name "Higgs boson" has become quite standard.

The problem with this particle was not in the name, but in the mass. The Weinberg-Salam theory could only predict the existence of such a particle, but not its mass. So there was no clue to the experimentalists where to look for this particle. This caused a lot of anxiety and a lot of soul-searching to find alternative mechanisms for explaining the masses of the W and the Z particles until in 2012 evidences of this particle were seen at the same place where the W and the Z were seen. The evidence is not conclusive yet: for example, it is not known whether the particle seen is indeed spinless. Scientists are waiting to obtain more data to settle the question. However, to the extent that the properties of this particle are known, it does look like the particle predicted in the Weinberg-Salam model.

5.9   The unifying hand

This is a wonderful situation. Three of the four interactions can now be explained by their corresponding gauge symmetries. It would really be nice to have the fourth one explained too. Sadly, gravitation still eludes us.

Yet, the physicists would not stop dreaming about a grand unification. Instead of four different kinds of interaction would it not be possible to explain everything by one single force, one single interaction?

Is that even conceivable? Why ever not? Even the electricity and the magnetism were thought to be two completely different phenomena till about 150 years ago. People thought that the force applied by one charge upon another has no connection whatsoever with the process of attraction of a piece of iron by a magnet. Now we know that it is not so. Maxwell has taught us that both of these arise basically from the same kind of interaction. Surely we can think of a single interaction that would unify the electromagnetic, the strong and the weak force, even if we have to keep the gravitational interaction aside for the time being.

We already know that the quarks respond to the strong force, but not the leptons. If there exists such an inherent difference between these two interactions then how would a unification be possible?

Of course, a difference does not mean there exists a contradiction. A little while ago we had been talking about hidden symmetries. In the context of magnets we said that there is no inherent difference between the direction along which a magnet's axis is aligned and any other. Perhaps we do see some difference, but that is only apparent.

It is conceivable that the quarks and leptons do not have any inherent difference after all. The difference lies only in their apparent behaviour. This was first proposed in two research articles published in 1974. The authors of one of these were Jogesh Pati and Abdus Salam. The other was written by Howard Georgi and Sheldon Glashow. The articles laid down two different arguments. Georgi and Glashow claimed that there is no problem with a grand unification. All the three interactions can be explained within a single framework. They also discussed the nature of the theory which would describe this unification.

There is a bit of unease, though. The strong force is extremely strong, even the electromagnetic force is much stronger than the weak force. How could they be described by a single interaction?

But what exactly do we mean by saying that the 'electromagnetic force is strong'? The electromagnetic interaction arises due to the exchange of photons. At any moment the probability of an electron emitting a photon determines how many photons would be emitted. The higher the probability, the larger the number of photons and stronger is the electromagnetic interaction.

It was observed that the interaction between two electrons would depend on their relative momentum. If the momentum of one is increased with respect to the other, the probability of photon emission would also increase, thereby increasing the magnitude of the electromagnetic force. However, the opposite happens in the case of gluon exchange. The probability of emitting or absorbing gluons decreases if the relative momentum of two quarks increases. Slowly, but surely.

For the energies (or, equivalently, momenta) achieved by particles in modern accelerators, the magnitude of the strong force is much larger than the electromagnetic force. But at higher momenta the strong force would progressively weaken and the electromagnetic force would become stronger. Consequently, they would become equal at some point and there would be no difference between the two forces. The weak force might also become equal to these two at this point. So all three forces would become one now and we shall have the grand unification.

This theory was proposed by Howard Georgi, Helen Quinn, and Steven Weinberg. Now the real problem lies in verifying this theory.

The momentum at which all three forces become equal is absolutely enormous. The best of our present day accelerators can generate particles with energies up to millions of MeV. To give an idea as to how large such energies are, it would suffice to say that the electrons inside an atom typically have kinetic energies which are about one hundred millionth of this. But if we could accelerate the particles such that they have energies that are millions of billions times as large as a million MeV, the unification of the forces could be observed. But there is no possibility of reaching such energy scales through human technology in the foreseeable future.

But would there be no indirect evidence for this grand unified theory? Yes, of course. For example, consider the charges of the electron and the proton. The magnitudes are equal. What is the reason for this?

This is not a silly question. When we said that the rest mass of a proton is equal to 1840 times that of an electron, it was a rough estimate. The precise ratio is somewhat like 1836.15152. There does not appear to be anything interesting here. The interesting coincidence is in the spin: because both have exactly the same spin. But then, according to the theory of quantum mechanics, the spin of a particle cannot be arbitrary. It can only be 0 or $\frac{1}{2}$ or 1 or $\frac{3}{2}$ and so on. Therefore, it is not very surprising for two particles to have the same spin. But it is not so for the case of the electric charge. A particle can have any amount of charge. Yet, the proton and the electron have the same amount of charge. The magnitudes are precisely equal, there is absolutely no difference. The problem lies here. The proton and the electron are totally different particles. Still, why do their charges have exactly the same magnitude? Or, we could think of this in another way. Why the charge of the d quark is exactly one third of that of the electron and not something else?

The answer came from the grand unified theory. We said that quarks can have three different colors. If the charges of the three d quarks of three different colors are added, then according to Georgi and Glashow, that would be equal to the sum of charges of an electron and an electron-neutrino.

Since the electron-neutrino is charge neutral, the charge of the d quark has to be equal to one third of the electron charge. There could be no deviation from this. The fact that the charge of a u quark has to be larger than that of the d quark by an amount equal to the proton charge Q was already known from Weinberg-Salam theory.

In the previous chapter we have discussed the conservation of baryon number, and of lepton number. Why are they conserved? The reason behind the conservation of charge is a gauge symmetry. Is it possible that there exist some such symmetries behind these conservations? Unfortunately, we do not know of any such symmetry. And yet, why don't we see any violation of baryon number or lepton number?

The grand unified theory provided an answer to this question. The exchange of certain particles can change the baryon number. But these particles are extremely heavy, the rest masses being at least 10^{13} times larger than that of W or Z. The probability of exchanging such a particle is therefore extremely low. Consequently, baryon number changing processes are exceptionally rare.

As far as we know, protons are stable particles. Why are they stable? We said that the basic reason is the conservation of baryon number. The baryon number of a proton is 1. And all particles lighter than a proton have zero baryon numbers. If the baryon number is conserved, a proton cannot decay.

But the grand unified theory says that there is no conservation of baryon or lepton numbers. Therefore, it is not impossible for a proton to decay. For example, there could be a process like,

$$\text{proton} \rightarrow \text{positron} + \text{uncharged pion}.$$

The charge of the positron is equal to that of the proton. Hence charge is conserved. The baryon number is 1 to begin with and zero in the end. The lepton number becomes -1 in the end, whereas it started with 0. So, none of these is being conserved.

Now that is a scary concept! Our bodies are primarily made up of protons. If they start decaying we should soon disappear like a puff of smoke. The same would happen to everything around us! But there is nothing to worry about really. Even if the proton is not absolutely stable, it is quite certain that it has a very long life. The average lifetime of a proton is at least more than 10^{32} years.

Just think about this — 10^{32} years! Compare this with the average life expectancy of a human being. In fact, let us say we wish everyone to become a centenarian, which is to have a lifespan of 10^2 years. The age of our Earth is a few billion years : that is 10^9 years. The age of the entire Universe is definitely less than 20 billion or 2×10^{10} years. The average life of proton is evidently far larger than any of these numbers. So there is nothing to worry about, really. The day of significant proton decay is very far in the future. None of us would be there to see that. It is not even clear if the Universe itself would exist in the form we know it to be now.

But all this is being said about the 'average' life of proton. This means that some protons would decay fast, and some would take much longer. And the average would come to about 10^{32} years. This means that a few protons would surely decay over a short time scale, within our own lifetimes! If that happens it would prove the grand unification theory conclusively.

Large laboratories have been built for this purpose. And the scientists are on constant vigil there to detect a proton decay event. Of course, nothing has been detected so far. But nobody can say that we would not observe such an event tomorrow itself. If and when that happens we would have conclusive proof of the grand unified theory, which can describe all three forces through a single interaction.

5.10   Untied knots, unexplored horizons

We hope that the alert reader has not missed an important point in all this. In the first chapter, when we talked about the classical physics the role of symmetry was somewhat like that of a brilliant supporting actor. We could have done without it under duress, but it was providing us with valuable support. Many questions could be answered far more easily than usual. After traversing a long road now we see that symmetry has acquired the central role in modern physics.

This is because today we are building up theories on the basis of various gauge symmetries which help us to understand the mysteries of nature. We have described the basics of the standard model of particle interactions. As we see, the model is based on gauge symmetries. In the case of weak interactions, the symmetry is hidden. For electromagnetic and strong interactions, the symmetry is apparent.

The model has been remarkably successful in describing particle phenomena. There are numerous situations where the predictions of the model can be calculated with high precision, and there the results of the experiments agree with the predictions of the model. Very roughly speaking, we have not seen any particle phenomenon which violates the basic tenets of this model. The only black spot in this wonderful success story is the discovery that neutrinos do have some rest mass. In the original version of the model, the neutrinos were assumed to be massless, since experimental data at that time were consistent with this assumption. However, the model can be modified without much trouble to accommodate neutrino masses.

That does not mean that all problems have been solved. There are quite a few open ends.

First, what happens to gravity? We do not know yet. It has not even been possible to describe gravity by quantum field theory. Unifying it with the other three interactions is a distant enough dream.

And we need not be too content about the unification of the other three interactions either. The decay of protons has not been observed yet. Is the grand unification not correct, after all? Or do we need to introduce modifications to it?

Perhaps we do. Because certain experiments, conducted after 1990, have measured the magnitudes of the electromagnetic and the weak forces very accurately. From these measurements it can be seen that at very large momentum the forces become quite close to each other but they do not become exactly equal. That is we do not have a perfect unity.

Of course, this is not a very serious issue. Because to estimate the magnitudes of these forces at very large particle momenta we need to make certain assumptions. One of these is rather important. According to grand unification a number of new particles would appear at the energy where unification is expected to happen. But that energy is so much larger than the energies being achieved at the present day accelerators. But would there be no other particles between the unification energy and the energy scales of today? Georgi and Glashow assumed that there would not be. There is no reason for that to be true. There have been many modified versions of the unification theory which predict a number of particles with energies in between. If that happens then the calculations have to be redone. The forces would perhaps unify at even higher energy. Perhaps the actual theory is more complex. And the results of recent experiments do indeed point towards such complexity.

Many other unresolved issues remain, besides the fundamental question of unification. We should recall the famous question of Rabi. What is the necessity of having second or third generation of particles? The universe could very well have been made using particles only from the first generation. Then why do we have muon or tau, why ever the c, s, t, b quarks? Who knows?

The theory of strong interactions bases itself entirely on the idea of the existence of quarks. This idea has explained so many experimental observations that it is hard to disbelieve it. And yet, it has to be remembered that it has not passed the acid test for any theory or any idea: no one has observed a quark in an experiment.

The reason for this might be that the quarks cannot be freed: they are perennially in bound states which are hadrons. Such things are not unheard of. One pole of a magnet cannot be freed from the other, the poles always come in pairs. Perhaps something similar happens for quarks. Well, perhaps, but that is a speculation. No one has shown that quantum chromodynamics leads us to this conclusion.

To a large extent, the problem lies in the fact that it is very difficult to calculate the effects of strong interactions when the quarks are far apart. Here the word 'far' must of course be taken in context: even the average distance of quarks in a proton would be considered 'far'. If one wants to free a quark by pulling it apart from a hadron, one has to pull it to even larger distances, where calculations are even more difficult and less reliable. The reason for such state of affairs lies in the strength of the interaction. For weak and electromagnetic interactions, more complicated Feynman diagrams for a process always give a much smaller contribution compared to the simplest ones. For example, consider the interaction between two electrons, mediated by photons. We showed some complicated diagrams in Fig. 5.3, and a very simple diagram, with only one photon exchange, in Fig. 5.2. But, unless one is worried about very minute corrections, the simplest diagram is all we need. And, even if one is worried about some minute corrections, one has to calculate only a few complicated diagrams, depending on the degree of minuteness that one is interested in. For strong interactions, such rules of thumb do not exist. Numerical calculations, not dependent on Feynman diagrams, can be performed on computers, but they have to make drastic compromises in the nature of the problem in order to reduce the problem to a calculable form.

There are aesthetic problems as well. The standard model, as it is, contains nineteen parameters. These parameters cannot be calculated: they have to be determined through experiments. With them as inputs, we can find the results of other questions. But the number 19 does not make one feel very comfortable. The number grows once one has to accommodate neutrino masses. If you have to give so many inputs to a theory, it leaves you with a creepy feeling that perhaps you are missing some deeper understanding which could have cut down on the number of inputs.

Indeed, one of the persistent dreams of physicists is the idea of unification. This is the underlying belief that we will not need different theories for different interactions: one theory will be able to describe all of them. There have been several suggestions regarding this dream, but no experimental confirmation for any of them.

Of course we do not know for sure whether Nature works on a unified scheme. But we know for sure that something is obviously missing in the standard model. At the very beginning, we said that we will not consider gravitational interactions because it is negligible at the scale of elementary particles. While it is true, it is also true that we do not know how to describe gravitation in the form of a quantum theory. Since the 1980s, string theories have raised the hopes of describing gravitation. It is not clear how and whether the other interactions are contained in such theories.

So there are a lot of things to be done, a lot of ground to cover. We have to walk a long way still. What's more, we do not even know whether there is an end of the road. Reflecting on the history of science, we see that whenever we have solved a mystery at a certain level, new mysteries at a new level have been exposed in front of us. Perhaps the journey is endless, and that is the beauty of the challenge.

Suggested reading

The following list was not a part of the original Bengali book. It has been added at the request of the publishers of the present edition. The publisher's name and other information about the books have not been mentioned in the list for two reasons. First, most (if not all) of these books have multiple editions. Second, these days such information is easily available from the Internet, and any interested person can find out the most recent editions at any given time.

1. Albert Einstein, Leopold Infeld • *The Evolution of Physics* (First published 1938). A lively exposition of classical physics and relativity theory.

2. Geroge Gamow • *Mr Tompkins in paperback* (Collection of two books, *Mr Tompkins in wonderland*, First published 1940, and *Mr Tompkins explores the atom*, First published 1945). A fictional journey which exposes the basic ideas of relativity and quantum theory.

3. V. Rydnik • *ABC's of Quantum Mechanics* (First published c. 1963). A very detailed and very readable account of the evolution and the successes of quantum mechanics.

4. Richard Feynman • *The character of physical law* (First published 1965). A fascinating account of the laws of classical and quantum physics

5. George Gamow • *Thirty years that shook physics* (First published 1966). An excellent personalized account of the birth of quantum physics.

6. L. Ponomarev • *In quest of the quantum* (First published 1973).

7. Gerald Feinberg • *What is the world made of?: Atoms, leptons, quarks, and other tantalizing particles* (First published 1977). A beautiful account of atomic and particle physics up to the time of publication of the book.

8. Heinz Pagels • *The Cosmic Code* (First published 1982).

9. Richard Feynman • *QED: The strange theory of light and matter* (First published 1985). First-hand account of one of the biggest architects of the quantum theory of electrodynamics.

10. Steven Weinberg • *Dreams for a final theory* (First published 1992).

11. Bruce A. Schumm • *Deep Down Things: The Breathtaking Beauty of Particle Physics* (First published 2004).

12. A. Zee • *Fearful Symmetry: The Search for Beauty in Modern Physics* (First published 1986).

13. Leon Lederman, Dick Teresi • *The God Particle: If the Universe Is the Answer, What Is the Question?* (First published 1993). The book that attached the unfortunate tag "God particle" to the Higgs boson.

14. Harald Fritzsch • *Elementary Particles: Building Blocks of Matter* (First published 2005).

15. G. Venkataraman • *Quantum Revolution I - The Breakthrough* (First published 1994).

16. G. Venkataraman • *QED: The Jewel of Physics* (First published 1994).

17. Yuval Ne'eman, Yoram Kirsh • *The particle hunters* (2nd edition 1996). A journey through the experiments that discovered elementary particles.

18. Lincoln Wolfenstein, João P. Silva • *Exploring fundamental particles* (First published 2011). Intended for an audience with some background in Physics.

Index

The numbers mentioned against each word indicate the section and the chapter of this book where the word has been used. For example, 3($\underline{4}$,6) mean that the corresponding word can be found in Sec.4 and Sec.6 of Ch.3. In particular, the underlined number signifies that a rough idea about the meaning of the word would be available in Sec.4.

Scientists

Planck, Max 2(6–8) 3(6–8) 4(1)
    5(3)
Ptolemy 1(1)
Quinn, Helen 5(9)
Rabi, Isidor Isaac 4(5) 8(10)
Raleigh, Lord 2(6,8)
Reines, Frederick 4(3)
Richter, Burton 4(9)
Rubbia, Carlo 5(8)
Rutherford, Ernest 3(4–6)
    4(8–10)
Salam, Abdus 5(8)
Schrödinger, Erwin 2(9) 3(7,8)
    4(1)
Schwarz, Melvin 4(5)

Segre, Emilio 4(2)
Steinberger, Jack 4(5)
Stern, Otto 3(7)
Thomson, George 2(9) 5(4)
Thomson, Joseph 3(3)
Ting, Samuel 4(9)
Uhlenbeck, George 3(7)
Volta, Alessandro 1(7)
Weinberg, Steven 5(8,9)
Wien, Wilhelm 2(6,8)
Yang, Chen-Ning 5(5,7)
Yukawa, Hideki 4(4–6)
Young, Thomas 2(3,10)
Zweig, George 4(7,8)

Scientific terminology

accelerator 4(**6**,8) 5(9)
α-particle 3(4) 4(10)
angular momentum 1(**3**)
    3(7,8) 4(2,7) 5(6)
antiparticle 4(**2**,3,5–7,9,10)
    5(6,7)
atom 1(4,5) 2(1,9) 3(1–6,10)
    4(1–4,10) 5(1,3,8)
atomic weight()
baryon 4(**6**,7)
baryon number 4(**7**,9) 5(9)
betatron 1(9)
broken symmetry 5(8)

color of quarks 5(**6**,9,10)
component 1(**2**,3) 3(7,8) 4(2)
    5(4,6)
combustion 1(5)
compound 1(**5**) 3(1,2)
conduction 2(6)
conservation of angular
    momentum 1(**3**,7,10) 4(3,5)
    5(4)
convection 2(6)
conservation law 1(**1**–7,9,10)
    4(3,5,7,9,10) 5(4)
conservation of electric
    charge 1(**4**) 4(3) 5(3,4,9)